The Swamp Peddlers

ALSO BY JASON VUIC

*The Yucks: Two Years in Tampa with the Losingest Team in NFL History*
*The Sarajevo Olympics: A History of the 1984 Winter Games*
*The Yugo: The Rise and Fall of the Worst Car in History*

# The Swamp Peddlers

## *How Lot Sellers, Land Scammers, and Retirees Built Modern Florida and Transformed the American Dream*

Jason Vuic

The University of North Carolina Press CHAPEL HILL

*This book was published with the assistance of the Wells Fargo Fund for Excellence of the University of North Carolina Press.*


Set in Merope Basic by Westchester Publishing Services
Manufactured in the United States of America

The University of North Carolina Press has been a member of the Green Press Initiative since 2003.

Library of Congress Cataloging-in-Publication Data
Names: Vuic, Jason, 1972– author.
Title: The swamp peddlers : how lot sellers, land scammers, and retirees built modern Florida and transformed the American dream / Jason Vuic.
Description: Chapel Hill : University of North Carolina Press, 2021. | Includes bibliographical references and index.
Identifiers: LCCN 2020051337 | ISBN 9781469663159 (cloth ; alk. paper) | ISBN 9781469663333 (paperback ; alk. paper) | ISBN 9781469663166 (ebook)
Subjects: LCSH: Real estate business—Corrupt practices—Florida. | Mortgage loans—Corrupt practices—Florida. | Fraud—Florida. | Real estate development—Florida—History. | Retirees—Housing—Florida.
Classification: LCC HD266.F6 V85 2021 | DDC 333.3309759—dc23
LC record available at https://lccn.loc.gov/2020051337

Cover illustrations: Cape Coral brochure courtesy of Steve Fasnacht (https://flickr.com/photos/electrospark); *inset,* Port Charlotte model home courtesy of Charlotte County (Florida) Libraries and History Division.

*To Asher and Imogene*

*Contents*

# *Illustrations and Maps*

ILLUSTRATIONS

MAPS

[Florida has] a glorious tradition of hawking itself as shamelessly and profitably as is humanly possible. If someone sells you swampland, it's because someone sold swampland to their fathers, and perhaps even to their grandfathers before that. Maybe it's in our blood, or maybe it's just something in the water, but it is part of our heritage.

—CARL HIAASEN, *Miami Herald*

# The Swamp Peddlers

# Introduction
## *The Swamp Peddlers*

> [In Florida] . . . the future belonged to an eclectic collection of hucksters and boosters, risk takers and fortune makers who saw retirement communities and instant cities where others saw sand and swamp.
>
> —GARY R. MORMINO, *Land of Sunshine, State of Dreams*

Tough, stylish, sophisticated, in 1970 the talented Elizabeth Whitney was an award-winning journalist and the lone female news editor at the *St. Petersburg Times*. She covered real estate, having worked her way up, as a thirty-ish single mother, from what she called the "female tract" of food, fashion, and church reporting to what in those days was the male-only tract of business. But Whitney knew what she was doing. She had edited a large student newspaper as a college undergraduate, and her father had developed the 3,700-acre Ridge Manor community north of Tampa. It was her family's business. Lot sales, home sales, tax assessments, downtown redevelopment projects, suburban planned communities, retirement communities, and even nursing homes: these were her bailiwick, and Whitney covered property-related topics with aplomb, winning "best article" awards from at least a half dozen entities, including *Florida Realtor* magazine, the National Association of Real Estate Boards, and the Florida Women's Press Club, among others.

In early 1970, with the encouragement of her boss, Donald K. Baldwin, the president and chief editor of the *Times*, Whitney set out to win a Pulitzer, of all things, after she attended a meeting in Tampa of a state agency known as the Florida Land Sales Board (FLSB). Its job was to regulate what had to have been, apart from pollutive phosphate mining, the state's seediest industry: the high-pressure sale of homesites and raw, unimproved acreage sight unseen to tourists, foreigners, soldiers stationed abroad, and northerners—your quintessential swamp peddling. People had done it for decades, of course, perhaps even a century or more, but things kicked into a higher gear following World War II, when three developer-brothers from Miami—Elliott, Robert, and Frank Mackle—popularized an entirely new model of land sales by hawking lots in planned communities in Florida's

Seen here circa 1970, the Miami-based Mackle brothers (left to right, Elliott, Robert, and Frank Jr.) revolutionized the sale of homesites in Florida by offering customers easy-to-pay, $10 down, $10-a-month installment plans instead of mortgages. Courtesy of Marco Island Historical Society.

southern and central flatwoods on the installment plan, for $10 down and $10 a month.

As Whitney knew, the Mackles were honest brokers; they only sold buildable homesites in perhaps a dozen large communities, but the billion-dollar industry they created, the so-called installment land sales industry that came and went in the brief period between 1955 and 1975, spawned hundreds of competitors who sliced and diced and ultimately sold the state one lot at a time. They ranged from publicly traded "land giants"—massive corporations headquartered in Miami that owned hundreds of thousands of acres and built entire communities from scratch, including Port Charlotte, Port St. Lucie, Port Malabar, Cape Coral, North Port, Golden Gate, Deltona, Marco Island, Rotonda, Lehigh Acres, and Spring Hill, among others—to fly-by-night and sometimes fraudulent entities selling forty-nine lots or less.

These so-called forty-niners, Whitney learned, were shady lot-sales operations that deliberately sold fewer than fifty parcels a year, the minimum required by the FLSB for both registration and the disclosure of "property reports," which told buyers whether or not the parcel they were purchasing was, for example, seasonally inaccessible or located in a swamp. Yet, even for bigger companies, the sale of swampland was legal. And Whitney found that despite property reports which had disclaimers written in red that blared "THIS LAND IS SUBJECT TO PERIODIC FLOODING" or "THIS LAND HAS NO PHYSICAL OR LEGAL ACCESS," buyers couldn't be protected from themselves. They didn't read the reports or had been sweet-talked by a professional salesman at a call center in Miami or by one of the 600 or so compatriots sitting in cubicles in at least thirteen different centers citywide who made some 4.5 million unsolicited sales calls each year.[1]

Whitney couldn't believe it. She, like everyone else, had figured swamp peddling was a thing of the past, a relic of the 1920s when Florida had its last great boom. But she stopped short when she heard an FLSB official exclaim that the sale of "unusable land" should be illegal. "It wasn't illegal?" Whitney asked, incredulously. That seemed impossible, but as she looked into the story and dug through the agency's files, she found that for Florida's installment land sales industry, the marketing of swampland as something other than swampland was par for the course. The Mackles didn't do it, focusing instead on selling "high and dry" homesites in drab but still functional communities such as Spring Hill and Deltona, but other land giants, including the state's biggest developer of all, the Gulf American Corporation and its successor company, GAC Properties, did.

Then, as Whitney discovered, there were 150 or so smaller corporations that had banded together into a lobbying group called the Florida Land Association (FLA), which she examined in a seven-part investigative series called "The Swamp Peddlers" in May 1970. Her goal in writing the series, apart from winning a Pulitzer, was to encourage the state legislature, in its 1970 session ending in June, to ban the sale of unusable swamp land once and for all. Therefore, she had to hurry, so she investigated the FLA for almost a month, then sat in a tiny room at the *Times* with only a desk and a typewriter where she knocked out the 15,000-word series in just two and a half weeks. "One night I must have been especially weary," she wrote. "Intending to call the home of [an FLSB attorney] in Miami, instead I reached Morton Rothenberg," the lawyer who represented the FLA. Whitney apologized to Rothenberg and tried to explain, but the lawyer joked: "Honey, you didn't call the cops. You called the robbers."[2]

In all, Whitney estimated that by 1970 the FLA's member institutions had sold $128 million in land, an astonishing $850 million in today's terms, but just a small portion of the installment land sales industry as a whole. It seems so mundane, buying land on the installment plan, but this is how Florida was built; not the urban Florida of Tampa or Miami or the so-called other Florida of tiny agricultural towns stretching through the center of the state and into the Panhandle, but the sprawling, 1950s-era, planned communities in between, those cookie-cutter exurbs with no downtowns where every northerner's grandparents seemed to live. That, in fact, is what these communities were at first: homes for grandparents, entire Levittowns built not for commuters and families but for Yankee transplants and retirees.

Levittown, of course, was the famed planned community in Nassau County, New York, on Long Island, which brothers William and Alfred Levitt and the firm Levitt & Sons developed on some 4,000 acres of farmland beginning in 1947. Considered the "prototypical American suburb," Levittown, New York, was the first of three American Levittowns where the Levitts used prefabricated materials and assembly-line building techniques to put middle-class Americans into homes.[3] In Miami, the Mackles did the same thing. They were contemporaries of the Levitts, having built large subdivisions in and around the city through the late 1940s and in 1951 a bedroom community on Key Biscayne. However, by 1955, the Mackles had changed their approach from building low-cost homes in urban areas to planning and constructing Levittown-like retirement communities in (what were then) rural sections of the state, where land was cheap and plentiful but a lack of jobs meant only pensioners could live there.

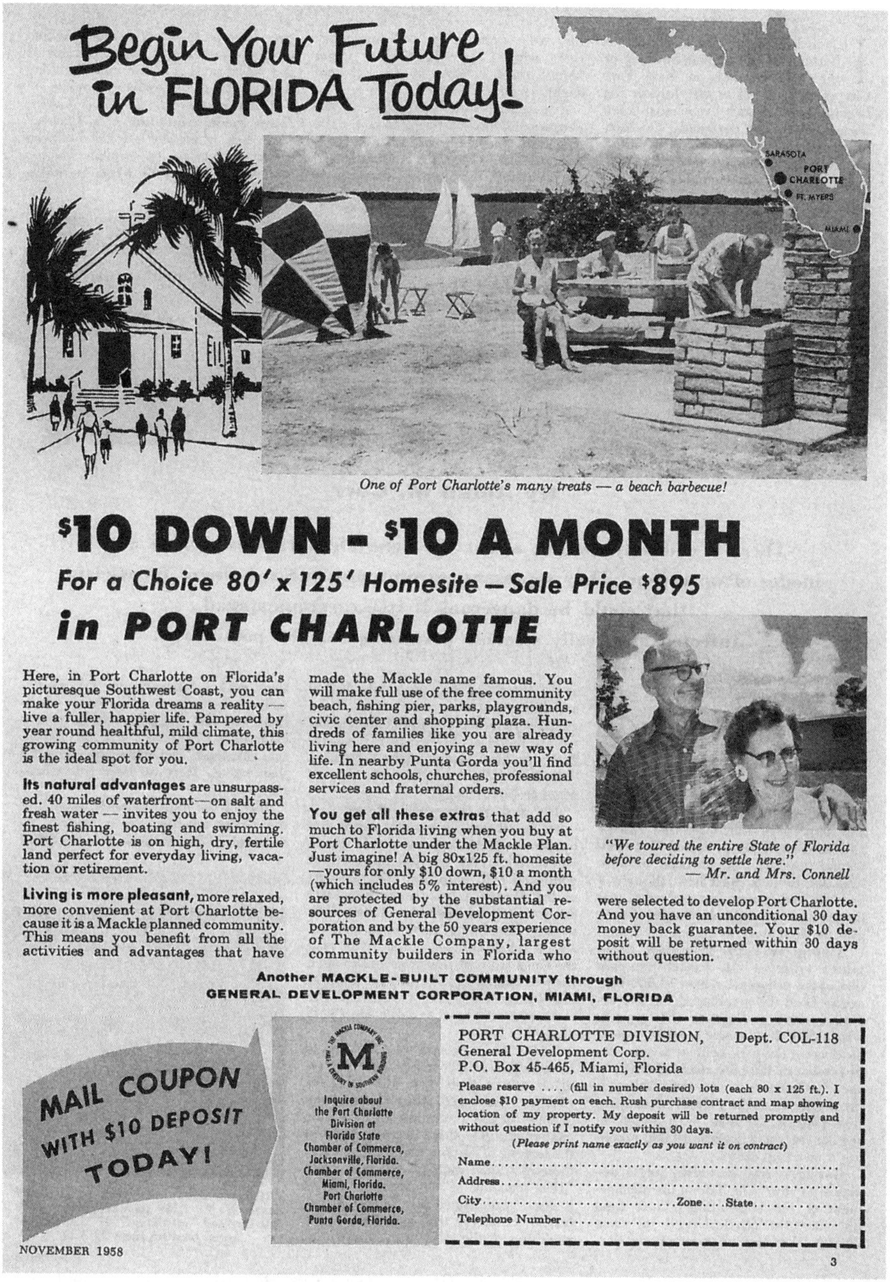

# Begin Your Future in FLORIDA Today!

SARASOTA
PORT CHARLOTTE
FT. MYERS
MIAMI

*One of Port Charlotte's many treats — a beach barbecue!*

## $10 DOWN - $10 A MONTH

### For a Choice 80′ x 125′ Homesite — Sale Price $895

## in PORT CHARLOTTE

Here, in Port Charlotte on Florida's picturesque Southwest Coast, you can make your Florida dreams a reality — live a fuller, happier life. Pampered by year round healthful, mild climate, this growing community of Port Charlotte is the ideal spot for you.

**Its natural advantages** are unsurpassed. 40 miles of waterfront—on salt and fresh water — invites you to enjoy the finest fishing, boating and swimming. Port Charlotte is on high, dry, fertile land perfect for everyday living, vacation or retirement.

**Living is more pleasant,** more relaxed, more convenient at Port Charlotte because it is a Mackle planned community. This means you benefit from all the activities and advantages that have made the Mackle name famous. You will make full use of the free community beach, fishing pier, parks, playgrounds, civic center and shopping plaza. Hundreds of families like you are already living here and enjoying a new way of life. In nearby Punta Gorda you'll find excellent schools, churches, professional services and fraternal orders.

**You get all these extras** that add so much to Florida living when you buy at Port Charlotte under the Mackle Plan. Just imagine! A big 80x125 ft. homesite —yours for only $10 down, $10 a month (which includes 5% interest). And you are protected by the substantial resources of General Development Corporation and by the 50 years experience of The Mackle Company, largest community builders in Florida who were selected to develop Port Charlotte. And you have an unconditional 30 day money back guarantee. Your $10 deposit will be returned within 30 days without question.

*"We toured the entire State of Florida before deciding to settle here."*
*— Mr. and Mrs. Connell*

**Another MACKLE-BUILT COMMUNITY through**
**GENERAL DEVELOPMENT CORPORATION, MIAMI, FLORIDA**

MAIL COUPON WITH $10 DEPOSIT TODAY!

THE MACKLE COMPANY INC.
M

Inquire about the Port Charlotte Division at Florida State Chamber of Commerce, Jacksonville, Florida. Chamber of Commerce, Miami, Florida. Port Charlotte Chamber of Commerce, Punta Gorda, Florida.

PORT CHARLOTTE DIVISION, Dept. COL-118
General Development Corp.
P.O. Box 45-465, Miami, Florida

Please reserve .... (fill in number desired) lots (each 80 x 125 ft.). I enclose $10 payment on each. Rush purchase contract and map showing location of my property. My deposit will be returned promptly and without question if I notify you within 30 days.

*(Please print name exactly as you want it on contract)*

Name........................................
Address........................................
City........................Zone....State..........
Telephone Number........................................

NOVEMBER 1958

3

To boost sales at Port Charlotte and its other "Port" communities, in the late 1950s the Mackle brothers/General Development partnership barraged northern markets with multimillion-dollar ad campaigns touting its $10 down, $10 a month installment programs. Courtesy of the author.

For $10 down and $10 a month—the price of "cigarette money," read one ad—retiring northerners could buy a piece of the Florida dream, an $895 graded homesite that would be waiting for them when they were ready to build, in this case, one of ten different home models, each similar in look and prodigiously spartan, constructed by the Mackles themselves. Yet, early on, the Mackles discovered that lot sales—not home sales—were the biggest and most lucrative part of the business. A 78,000-acre cattle ranch bought for $40 an acre could be chopped up into 200,000 lots, each 80′ × 125′—indeed, whole horizons of lots, with boulevards and side streets and access roads leading to nowhere. It was as if pop artist Andy Warhol had been a real estate developer, stamping out homesites instead of soup cans in a palmettoey boondock deep in the Florida bush.

So, what was the draw? Why did people buy property sight unseen, especially in Florida, a state with a history of land scams and more than 15 million acres of wetlands and swamps? It simply boggles the mind. The answer is Florida was, well, Florida—a place, wrote one author, where people didn't buy property but a "vision" of what they and their property would be, what they would someday be, with no more floors to shine, or buses to drive, in a new home in a new town that (a salesman assured them) was somewhere close to a beach.[4] In fact, the allure of Florida was so great that salesmen in northern cities actually sold lots from kiosks, like chewing gum. There were coupons in magazines. You filled them out and sent a check, and a contract arrived with a map marked with "X" where your lot was supposed to be. Other salesmen went door-to-door, made cold calls, held sales dinners, and led hundreds of bus and plane trips to Florida for those who wanted to buy but first wanted to see.

By 1959, the Mackles had sold some 65,000 lots.[5] Their first installment company, the General Development Corporation (GDC), had founded Port Charlotte, North Port, Port St. Lucie, and Port Malabar, among others, but faced numerous competitors, including the Gulf American Corporation, which, under the leadership of Jack and Leonard Rosen, two less-than-ethical former shampoo salesmen from Baltimore and the founders of Cape Coral, skirted the line between reputable land dealers and fraudulent peddlers of swamp. As Frank Mackle would say, "We started the ten-dollar-down, ten-dollar-a-month deal, but it drew in a lot of bad people who abused the business." These people included the Rosens, whose army of salesmen hoodwinked buyers into thinking a 55,000-acre stretch of the Everglades would yield buildable homesites when it wouldn't; Joe Klein, the head of Cavanagh Communities and its Rotonda West development, who publicly admitted, in a thirty-eight count settlement with the Federal Trade Commission, that the lots he had been selling were "of little value as investments and little use as homesites"; and E. John Wentland, whose Continental Land Management offered thousands of mostly retired investors a whopping 12 percent return on "first-mortgage bonds" backed by homesites Wentland didn't own.[6]

The scams were relentless, but the FLSB, Whitney knew, was a wishy-washy state agency that alternated at times between industry watchdog and industry shill, because the bottom line was that installment land sales brought a lot of money to the state, an estimated $2 billion (or $13 billion in today's terms) between 1955 and 1970. Former Florida governor Fuller Warren actually worked for a swamp peddler, as did a former state supreme court justice and two former FLSB directors.[7] Thus, though Whitney's 1970 "Swamp Peddlers" series was incredibly revealing, winning a distinguished journalism award from the Florida Society of Newspaper Editors, the state legislature wasn't going to act. It already required developers to issue extensive property reports, so if a person wanted to buy land, even swampland, as an "investment property," there was nothing more it could do.

And besides, even reputable land giants sold something that wasn't: homesites that were "high and dry" but had water in the driveway; "championship" golf courses with just nine holes; lots on drainage ditches that salesmen coolly pitched as "waterfront"; and backyard canals with docks for fishing that ruined the state's environment and killed its estuaries and actually were bad for the fish. With no exaggeration, Florida's land giant developers helped to terraform the state by dredging and filling and backhoeing and bulldozing what had mostly been wet pineland into millions of treeless rectangles, the fundamental building blocks of suburbs everywhere but in parts of Florida, an endless gridiron of brutally denuded 80′ × 125′ lots. "From Fort Lauderdale we flew west" across the state "to the Fort Myers area," wrote journalist Jacquin Sanders in 1985, "where [we] saw a most depressing vista: acre after acre [had] been cleared and subdivided into cramped suburban streets, before there [were] even people to live there. In all this acreage there [weren't] a dozen houses—only a grey, treeless landscape waiting for people to move in to justify the developers' optimism."[8]

Sanders had been looking for manatees; he'd been counting them from the air for a story he was working on, but instead he found Lehigh Acres, a former cattle-ranch-turned-lot-sales-development that had only 17,000 residents but an eye-popping 135,000 homesites, of which more than 121,500, or 90 percent, stood empty.[9] Like other land giant communities, the sprawling

Lehigh development had no downtown and no industrial zones, because the goal of its developers wasn't to create a community, per se, but to maximize the number of lots. They gave "no thought [as to] how communities would look," said one observer, or "how many lots you could assemble or how fast you could sell them," and certainly "no thought [as to] whether there'd be [enough] room there for commercial or industrial uses. . . . None of that existed prior to the mid-1970s."[10]

That's when high inflation rates and the advent of state and federal environmental laws and commonsense planning restrictions put the kibosh on installment land sales. The model no longer made sense financially, and communities that size just couldn't be built anymore. Those that were built on a smaller scale had to have better roads, water- and sewer-line hookups instead of septic tanks, and land set aside for parks and schools. They also had to preserve wetlands instead of destroying them, and dredging and filling, which had created land in the past but hurt the state environmentally, was now forbidden. And, though still not illegal, the selling of investment acreage, the uninhabitable swampland that Whitney wrote about, became harder and harder to do. "The old days of the very flashy margins in this industry are gone," said Herbert Kaplan, the president of the land giant Royal Palm Beach Colony, in 1973. "A developer today has to plan his community ahead of time. I think market conditions, as well as new regulations, require it."[11]

Yet, what about Lehigh Acres or Deltona or the other land giant communities, not to mention the dozens of investment acreage subdivisions (that were platted and sold but lie uninhabited today as so-called ghost subdivisions) that were grandfathered in? In other words, what has installment land sales, the strange and at times fraudulent industry that more closely resembled aluminum-siding sales or rare-metal sales than traditional real estate, meant for modern Florida? The answer is a lot, because even though relatively short-lived, the industry gave Florida maybe two dozen freestanding communities where today more than a million people live; by the year 2050, at buildout, these communities will be among the most populated areas of the state. North Port, for example, will have 265,000 people; Lehigh Acres, 350,000; and Cape Coral, 400,000—the latter spread out over a 120-square-mile territory bigger than Boston and San Francisco combined.

And because of decisions made in the 1950s, when single-family homesites were the cornerstone of Florida land development, virtually every one of these communities is destined to become—if it isn't already—sprawl. Because this is what happens when you divide Florida prematurely, with no rational planning, into a labyrinth of streets and millions of residential lots. You get pollution, traffic, poisoned and destroyed aquifers from relentless canal building and dredging and filling and runoff; low tax bases stretched thin for want of business districts or industries; and environmental damage on a mass scale with cleared lots now growing invasive melaleuca trees and Brazilian pepper trees instead of scrub brush and pine. This is the legacy of swamp peddling, when a generation of land developers, both good and bad (but mostly bad), subdivided a wet, low-lying, environmentally sensitive state for profit.

CHAPTER ONE

# $10 Down, $10 a Month!

> The idea from the beginning . . . was to sell lots, nothing more. No forethought was given to the possibility that someone might actually want to live there, no consideration given to streets or schools or sewers. . . . Selling, not building, was how these developers measured their success.
>
> —PAUL REYES, *Exiles in Eden*

They called it the "Ninety-Mile Prairie," the vast expanse of scrub brush and grassland stretching from south central Florida, above Lake Okeechobee, to a handful of ports on the Gulf. This was cattle land, a mucky, barren, rattlesnake-infested, palmetto-bushed terrain where in the 1800s grizzled cowboys, known locally as cow hunters, herded livestock onto steamships bound not for northern markets but for Cuba. Cows roamed freely here, malnourished and emaciated, while Florida's unique breed of cowboys, wrote Frederic Remington, had "none of the bilious fierceness and rearing plunge" of their Western brethren.[1] In fact, many had stolen whole herds—range wars were common in Florida—while some, such as cattle baron and slave owner Jacob Summerlin (who also cofounded Orlando) spent the Civil War selling "scrub cows" to both the Union and the Confederacy.[2]

In 1862, Summerlin and his associates established a dock on the north shore of Charlotte Harbor, several miles inland from the Gulf, for the express purpose of loading ships in secret to break the Union blockade. However, Summerlin's ships weren't going to the Confederacy because no one could pay what the Cubans could pay, a reported $30 a head, in gold Spanish doubloons.[3] Summerlin would die in 1893 after a long and prosperous life, but "Charlotte Harbor Town" lived on as the first settlement in what became Charlotte County. For years, it was nothing more than a remote hamlet on either side of an old Indian trail, the future U.S. Route 41, where settlers bought supplies and cowboys brought their livestock to the dock.

Across the water on the south shore of Charlotte Harbor was a jutting bit of pine and wetland called Punta Gorda, which was Spanish for "Fat Point." In the mid-1800s, Cuban fishermen sometimes visited there, but in 1883 the land passed to a Kentucky colonel and carpetbagger named Isaac Trabue, who platted a town with a port and persuaded the Florida Southern Railroad to

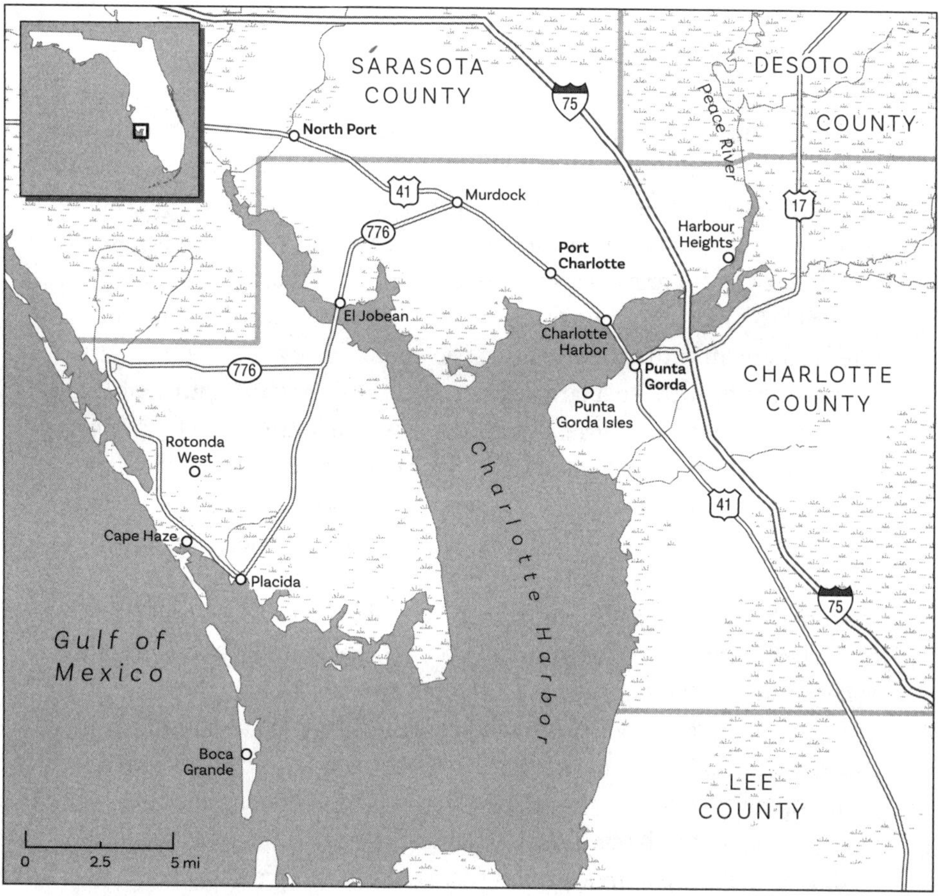


Charlotte County, Florida. Map created by Erin Greb Cartography.

build a terminal there in 1886. The railroad would become part of the "Plant System," owned by northern industrialist Henry Plant, who, in addition to building train and steamship lines from Canada to British Honduras, would open or acquire eight Florida hotels, including one built in 1887 in Punta Gorda.

Although not as luxurious as Plant's Tampa Bay Hotel, which was a quarter-mile-long, minaret-festooned showpiece featuring the state's first elevator, the Punta Gorda Hotel was an appealing though utilitarian structure of 150 rooms. Built to draw tourists to one of the best fishing grounds in the state, the hotel housed over 3,300 visitors in 1887, including such luminaries as Andrew Mellon, John Wanamaker, and W. K. Vanderbilt, among others, but closed following an economic depression in 1896 and didn't reopen until 1902.

The hotel and, to some extent, the communities of Charlotte Harbor never recovered. Some industry remained there, mostly fishing and phosphate shipping, but competition from ranchers in Texas and Central America meant that by the turn of the nineteenth century, the area's cattle export industry had dried up. Thus, with few tourists and even fewer exports to speak of, Charlotte Harbor spent the next fifty years in a stasis, for while the population of Florida grew 424 percent between 1900 and 1950, to nearly 2.8 million, Punta Gorda's grew just 37.8 percent, to 1,950. Charlotte Harbor Town (or simply Charlotte Harbor as the locals called it) remained an unincorporated settlement with fewer than 500 people.

Impeding growth, to a large extent, was the "wet-dry" nature of local lands. To the east was prairie, but north and south of Charlotte Harbor was a coastal strip of what scientists call "hydric pine flatwoods," a low-lying, lightly forested, seasonally flooded savannah of palmetto bushes, cabbage palms, and slash pine for which drainage of any kind was poor. It was so poor, in fact, that for much of the region's June through September rainy season, Southwest Florida was one big bog, receiving more rain in four months (35 inches, on average) than the damp Great Lakes states Wisconsin (32.6), Michigan (32.8), and Minnesota (27.3) received in a year. Rain, however, was one thing. Porous soil with high sand content, sitting on a shell-filled substrate, was another. This meant that instead of being absorbed, water pooled in the summer between one and two feet in places, then evaporated completely in the winter. By early spring, the land was dry enough for brush fires.

"It is amazing what dreary miles there are of these pine barrens," wrote one discomfited visitor in 1918, "arid level wastes with an undergrowth of palmetto scrub . . . and stretches of shallow bog pools. The roads are often under water in places," and "the utter absence of landmarks make[s] it unsafe for the stranger to wander from the beaten path."[4] Under these conditions, year-round agriculture was impossible, so companies used the land for logging of old growth cypress and mature pines and for the tapping of pines for resin. The latter practice was, in the 1800s, among Florida's largest industries: turpentine and pine tar production for hundreds of commercial uses, such as soap, paint, varnish, solvents, lacquers, furniture polish, lamp oil, and sealants.

It was a nasty business, in which freed slaves and other African Americans, many of whom were prisoners, would cut the bark off trees in three- and four-foot sections to score the wood underneath. This would "bleed" the trees of resin, which workers collected in buckets and took to so-called turpentine camps for distillation. There were several such camps north of

Although installment land sales companies promoted Florida as a tropical paradise, the lots they sold were typically located in hydric pine forests like this one, barren former cattle ranches, and sometimes swamps. Courtesy of Chris M. Morris/Flickr.

Charlotte Harbor, including one called Southland near the present-day community of El Jobean. "It was a horrible place," remembered one local resident. "The camp leased prisoners, all black. They were treated like slaves. Several were killed in beatings or died in forest fires from the pine resin dripping from the trees. . . . It was uncivilized."[5]

Not only that, but Florida's turpentine camps were incredibly destructive environmentally. Whole swathes of flatwoods were either bled out or cut down, leaving scarred, swampy, denuded, near-worthless wastelands in their place. That is, unless you could sell the land sight unseen to naive buyers, which is what Chicago real estate agent and land developer John M. Murdock did in 1912. Murdock's empire was a modest stretch of subdivided farmsteads approximately six miles north of Charlotte Harbor, which he advertised to northerners as a "vast area of fertile farms and gardens" where vegetables grew "every month of the year."[6]

Residents soon learned, however, that growth was impossible in the summer months, when the "Murdock Farms Colony" was underwater. Complaints proliferated and inhabitants left, which prompted a panicked Murdock to persuade county officials to construct more than a dozen large canals from the shores of Charlotte Harbor north to the county line. Unfortunately, the canals were too deep, causing a drop in the water table, which meant Murdock had exchanged too much water in the summer for not enough water in the winter. The colony collapsed, and in 1916, Murdock abandoned his debts and his family and fled with his secretary to Jacksonville.

Over the next few decades, the lands of north Charlotte County, including those of John M. Murdock and those of the turpentine companies, passed into the hands of a former railroad telegrapher named Arthur "A. C." Frizzell. An old Alabama farm boy with a taste for liquor and a talent for making money, Frizzell began in 1918 with twenty acres and a turpentine still, which he parlayed into a cattle and land empire of enormous size. The trick, he found, was in buying fallow land from lumber companies—land with few trees but thousands and thousands of stumps, which he "grubbed out" (using leased prison labor) then sold for their resin. "When the turpentine and lumber were gone," remembered a Frizzell family member, "the big companies wanted to get rid of the land so they wouldn't have to pay taxes on it." Therefore, they sold it to [A. C.] for fifty cents an acre. "In time, he had a lot of land with nothing but stumps and palmettos."[7]

As fate would have it, though, the stumps turned out to be a goldmine because they contained concentrated slivers of extremely valuable pine resin known as "fatwood." To get to the fatwood, Frizzell's men pulled stumps from the ground using a jerry-rigged car as a tractor. Then, they burned the stumps in a kiln—which, actually, was a hole they'd dug lined with sheet metal, covered in wood and sand for smoldering—and, as the stumps began to sweat, resin would drip through a pipe and into a bathtub. From there, it was poured into barrels and sold. The process was painstaking at best, which is why turpentine companies didn't do it, but old pine stumps made A. C. Frizzell a rich man.

He grew even richer during the Depression, when, as the stumps ran out, he planted his land with hardy Bahia grass, which grew well in Florida, and went into the cattle business. He had so many cows and so many ranch hands that at one point he began paying the men in "Frizzell dollars" good only at his Murdock store. When a bridge spanning Charlotte Harbor opened in 1921 and the Tamiami Trail from Tampa to Miami was completed in 1928, he went into the auto business, opening two Ford dealerships and a tractor dealership, plus a lumberyard. With upwards of 100,000 acres, by the 1940s Frizzell was one of the largest landowners in Florida, and on occasion he could be seen wearing a suit and a Stetson hat and being driven about in a Ford by a black assistant named Sam.

Frizzell lived well until his wife died in 1953 and he lost interest in the business, selling the bulk of his land, a reported 78,000 acres, for $3.6 million in 1954.[8] The buyer was a Canadian company called the Chemical Research Corporation, a one-time petroleum business that had bought so much land in Florida and become so Florida-focused that it changed its name to the Florida Canada Corporation. Why? Because this was the boom, the second great boom of frenzied land sales in Florida in the twentieth century. The first had come in the 1920s, when a bevy of dreamers, schemers, real estate agents, land developers, government officials, bankers, and even confidence men had driven the state's land prices into the stratosphere.

Theories as to why this happened abound, but suffice it to say, the Roaring Twenties were a decade-long period of sustained economic growth. For the first time, more Americans lived in cities than in the countryside. They were earning more than ever and retiring with healthy pensions more than ever, and thanks to mass-produced automobiles such as the Model T, they were mobile. They had "leisure time," a concept unknown to all but the richest Americans until well into the 1900s. And, with the completion of U.S. Route 1 from Maine to Miami in 1915 and the above-mentioned Tamiami Trail, they could spend it in Florida, or even live in Florida—where beginning in 1924 there was no state income tax—in some of the first modern planned communities in the United States.

Unlike in earlier generations, when Henry Plant and other industrialists built perhaps two dozen hotel-spas in Florida for America's super elite, who were tourists, these new communities were year-round and decidedly middle class. In 1925, for example, Cleveland's Brotherhood of Locomotive Engineers developed a 2,900-acre stretch of Venice, Florida, as a Mediterranean-style retirement community for its workers. Likewise, the settlement of El Jobean in northwest Charlotte County began as a defunct turpentine camp purchased in 1923 by Boston land developer Joel Bean—hence the anagram El Jobean—who mapped out six hexagonal "wards," each with a one-hundred-foot-wide circular boulevard with a civic center in the middle. In both instances, huge tracts of farmland, cattle land, orange groves, and/or scrub brush were drawn up into subdivisions, officially platted, broken up into thousands of half-acre, quarter-acre, and 80′ × 125′ lots, then sold.

"Streets are rapidly being marked and graded," exclaimed an ad by the promoters of muddy Clewiston, Florida, in 1926, and "sidewalks [are being] laid and waterlines installed. Water supply pumps and generators are in use. Gigantic dredges and earth moving machines are rapidly completing the undertaking specified by Clewiston's city planner, Clewiston's architects, engineers, and landscaping authorities . . . men whose records show that they are skilled in the development of permanent communities. . . . Clewiston is going forward! Do not wait until it has passed beyond your reach."[9] Built literally from nothing on the southeast shore of Lake Okeechobee by a Tampa banker and project cofounder named Alonzo Clewis, the proposed city of Clewiston offered homesites for $900, with 25 percent due at purchase and the rest over a set period at a fixed rate of interest.

But, as Florida boomed in the 1920s, buyers throughout the state, especially in Miami, found they could flip properties for astonishing profits before a single monthly payment was due. These were the so-called binder boys: buyers of subdivided lots who purchased options to property for a down payment of 5 or 10 percent, known as a "binder," which they immediately sold, often to other "binder boys," at a markup. The pace was furious; the speculation, rampant, "but few people worried much about the payments which were to come," wrote *Harper's* editor Frederick Lewis Allen in 1931, as "nine buyers out of ten bought their lots with only one idea, to resell, and hoped to pass along their binders . . . at a neat profit."[10]

In Miami, the absolute epicenter of Florida's land craze, fifty to seventy-five Pullman passenger cars arrived at the city's train station every day. They were filled with northerners: greedy, gullible, get-rich-quick investment seekers who were dying to get in—so much so that in the summer of 1925 alone, Miami property values rose 580 percent. That year, there were over 970 new local developments, including one in the Arch Creek section of Miami Shores that sold $33 million in property in three hours. For the year, Coral Gables did $100 million in sales; Miami Shores, $75 million; Miami Beach, $23.5 million.[11] (That's property in just three developments, in 2021 terms, selling for over $2.5 billion.)

"The boom seems almost too bizarre to have happened," opined one author, but "in retrospect . . . it was one of those strange spasms of speculative madness that occasionally seizes mankind."[12] It seized Henry Ford, of all people, who developed property in the scrub brush settlement of LaBelle, near Fort Myers; John Ringling in Sarasota; and even William Jennings Bryan, the famed orator and politician, who gave impassioned speeches on the virtues of "Florida living" to hordes of bussed-in land buyers in Coral Gables. Soon everyone had property everywhere—all of it overpriced and much of it underwater.

With so much land in play and prices in supposedly "good" developments so ridiculously high, by early 1926 sales evaporated: "We just ran out of suckers," said one Florida promoter—buyers defaulted and banks foreclosed.[13] Then, in September 1926, the most expensive hurricane to ever hit the United States ravaged southern Florida, a Category 4 monster that left some 370 people dead and 43,000 homeless in Dade County alone. (The hurricane also went inland, where it smashed levies at Moore Haven, a planned community near Clewiston, and killed 150 more.)[14] With that, the first Florida land boom was over. The state languished and cities that had overextended themselves by borrowing money for roads and water mains to once-booming subdivisions went bust. In fact, formerly wealthy St. Petersburg reportedly was the most indebted city per resident in the United States.[15]

Other rural developments disappeared, becoming ghost towns with one or two settlers and hundreds of vacant lots: Olympia, Brighton, Sparkman, Chevalier, Aladdin City, Winter Beach, El Jobean, and so on. A typical scene in Florida was the road to nowhere, named by a developer for his friends or family members, for U.S. states or flowers, or for whatever theme he was selling at the time. The Aladdin Company of Bay City, Michigan, for example, a mail-order home manufacturer that had staked out a Moorish-themed development fifteen miles north of Miami, gave us Hassan Street, Sinbad Street, Bagdad Avenue, and Ali Baba Circle. Likewise, in developer Joel Bean's El Jobean—the "City of Destiny" he called it—the streets were named for Bean's co-investors. He would die there penniless in 1942, having laid out two or three wards, but homes for just thirty-six people.

The hits kept coming. In 1928, Florida endured the deadliest hurricane in its history, a second Category 4 that walloped Palm Beach County and drowned some 2,500 people, mostly African-American farmworkers, in the floodwaters of Lake Okeechobee. (This was the storm immortalized in Zora Neale Hurston's classic work *Their Eyes Were Watching God*.) In April 1929, a Mediterranean fruit fly infestation led to a National Guard quarantine of all citrus from Florida, which devastated the state financially, then in October 1929 came the New York stock market crash that started the Great Depression. Yet Florida had already been through a depression, for three long years. "We'd gone through the collapse," recalled one Florida resident, and we'd "suffered the ravages" of the hurricane. "So when the stock market crashed" in 1929 "we were at the bottom of the well."[16]

What brought them out was World War II, when the U.S. military built or expanded over 200 installations in Florida and 2.1 million soldiers—approximately 14 percent of the near 15 million men and women who served in the war—came there for training.[17] They were joined by hundreds of thousands of workers, who, while building ships or barracks or testing airplanes, effectively kick-started the state's economy. In Starke, Florida, for example, a tiny, rural community between Gainesville and Jacksonville, the U.S. Army constructed the 170,000-acre Camp Blanding, which by 1942 housed some 60,000 people. By population, it was the fourth-largest city in Florida.

With so many people supporting the war effort, Florida had a housing shortage, prompting the military to build temporary barracks for soldiers and workers, known as "hutments," and to requisition entire hotels (such as the Don CeSar in St. Petersburg, which became an Army Air Forces convalescent center) and 85 percent of the hotel rooms, a reported one-fourth of Florida's entire hotel space, on Miami Beach.[18] Naturally, soldiers meant nightlife; they also meant gambling and drinking and taking part in all that the "Sunshine State" had to offer. And, as friends and family members visited, so did tourists, who in spite of food and gas rationing and a limit on train cars, came to the state in droves.

"If the fighting men of Anzio and Kwajalein could see Miami and Miami Beach now," wrote *Life* magazine in a scathing article in February 1944, "the anger they bear toward their enemies would probably turn against their countrymen," because "this Florida area has been indolently sunning itself in the brazen rays of a civilian boom that seems like something from the giddy past." As proof, *Life* showed pictures of what it called "acres of automobiles" in the parking lot of the Hialeah Park race track, where visitors bet $800,000 in a day but volunteers at the "Buy War Bonds" booth didn't sell as single bond.[19] A year later, while the war was still raging, *Life* visited Miami again, this time to examine real estate sales that had reached 1920s proportions, with some properties having "been bought and sold six times in a year."[20]

For millions and millions of Americans—be they tourists, workers, or pensioners—Florida was the place to be, only now the war had introduced the state to an entire generation of soon-to-be-discharged soldiers. They had to go somewhere, so after the war many returned to Florida. "I've been to a lot of places since I got in the Army," said a sergeant from Texas in 1945, "and this is where I want to live."[21] Unfortunately, prices for preexisting homes in urban areas were high, and there weren't enough homes, so as World War II ended, builders turned their attention to the suburbs, where subdivisions were small, at first, and homebuilding was piecemeal.

The demand, however, due in part to the easy availability of government-subsidized Federal Housing Administration (FHA) and Veterans Administration (VA) loans, was so intense that cities such as Jacksonville and St. Petersburg began freeing up property by giving away tax-delinquent lots, at little or no charge, to veterans. In this way, officials in St. Petersburg, where the cost of new residential building set records every year between 1946 and 1950, successfully reignited a failed 1920s subdivision called Rio Vista.[22] Nevertheless, as sales rose and moderately priced suburban land became harder and harder to come by, developers looked to what the *New York Times* called "rural and small town areas in off beat sections of the state," such as Charlotte County.[23]

There was the 1954 Chemical Research Corporation purchase of the 78,000-acre Frizzell Ranch, which followed the sales in 1952 and 1953, respectively, of 18,000 acres of rural Lee County east of Fort Myers to Chicago businessman Lee Ratner and 294,000 acres in Collier, Lee, and Hendry counties near Naples to the Gerry family of New York.[24] Other notable sales included 25,000 acres of scrub brush near El Jobean to William and Alfred Vanderbilt in 1951; 8,500 acres of ranchland in St. Lucie County, north of West Palm Beach, to media magnate Gardner Cowles in 1952; and 12,000 acres of rural Hillsborough County south of Tampa to the Universal Marion Corporation in 1956. Of course, multimillion-dollar land deals in the interior of Florida were highly speculative, but at $14 an acre, which is what the Gerry family paid, for example, buyers could flip the land for a profit. (In fact, the Gerrys would sell about a seventh of their holdings, over 42,000 acres of semi-submerged swampland in eastern Collier County, for $129 an acre in 1962.)[25]

Other investors, however, such as the Chemical Research Corporation, pursued a different tack, building from scratch entire planned communities by dividing the acreage into tens if not hundreds of thousands of lots. Then, they'd sell the lots through an inexpensive though financially dubious sales instrument called a "contract for deed," in which the purchase of the property was financed not by a bank or a credit union but by the corporation itself. Thus, there were no mortgages, only payments made, often at $10 down and $10 a month for the next eighty months. In exchange, after their final payment, buyers received—at least in theory—a ready-to-build lot, a "homesite," in a completely new community with roads, bridges, power lines, drainage systems, and perhaps even a golf course preinstalled.

In this way, millions of lower- and middle-class Americans, usually northerners in their fifties nearing retirement age, began the process of moving to Florida *one installment at a time*. Of course, companies had sold property using "contracts for deed" before, especially during the 1920s, but no one had ever done it on such a gargantuan scale. The Frizzell Ranch, for example, would be diced up and sold off in the form of 200,000 lots, each sized 80′ × 125′ and accessible by a paved road.[26] Likewise, in just five years, the Gulf American Corporation would elevate its homesites in nearby Cape Coral to a septic-tank-necessary 5½ feet above sea level by dredging millions of tons of dirt from 168 miles of canals, three yacht basins, and fourteen artificial lakes. As one scholar noted, workers collected enough dirt each week for an area the size of a football field fifty feet high.[27]

The costs were astronomical, but installment land sales on a mass scale benefited developers in two key ways. The first was the price, as in 1958 the cheapest lots at the former Frizzell Ranch went for $895 apiece. With roughly four lots to an acre, the Chemical Research Corporation stood to make $3,580 an acre for land for which it had paid less than $44 an acre. Second, and perhaps most importantly, by selling land on an installment plan, developers passed the finance costs of contractually obligated land improvements to the buyer, who each month dutifully mailed the developer a check. If buyers didn't mail the check, under the terms of the contract they lost all claims to the land and the monies they had paid were forfeited. For the developer it was an incredible deal, provided interest rates and inflation rates remained stable and lots could be sold en masse.

Therefore, developers cast a wide net, using radio, TV, and magazine ads, mail-order coupons, "win a week" in Florida contests, game-show giveaways, film showings, sales dinners, bus tours, celebrity "spokespitches," and phone calls—anything they could to do to bring in the checks. "If we could get enough land sold," said one executive, "we could do all we wanted to do" in terms of development.[28] That is, if they knew how to develop, because some companies, like the Chemical Research Corporation, didn't. It had money and land, and though it changed its name to the Florida Canada Corporation in 1954 and opened a subsidiary called the Florida West Coast Land Company, which purchased the Frizzell Ranch, it had no experience turning scrub brush into lots, let alone communities. For that, it turned to the best Florida developers in the business: Elliott, Robert, and Frank Mackle, who, in less than a decade, had built an astonishing 10,000 homes in six different subdivisions near Miami and Fort Lauderdale and opened the well-known island community of Key Biscayne. In 1955, the Mackles came to an arrangement with Florida Canada in which they agreed to turn the barren Frizzell Ranch into the planned community of "Port Charlotte."[29]

DUBBED THE "SELLERS OF SUNSHINE" by one magazine, the Mackles were tall, broad-shouldered family men who quite literally were born to the business.[30] In 1908, their father, Frank Sr., an English immigrant, had opened a Jacksonville construction firm that built dozens of large buildings in the South, including department stores, bank buildings, warehouses, and tenements, plus a hotel, a girls' school, a thirty-eight-building Army repair shop at Fort McPherson, Georgia, and a massive hydroelectric substation that powered Atlanta's street cars. Backed by ambitious advertising—a Mackle hallmark—the family rose to prominence primarily in Atlanta, where matriarch Theresa was a headstrong socialite who sponsored, among other things, the city's "perambulator parade." Frank Sr. hosted the "Mackle Cup," Atlanta's amateur golf championship, and sent his three sons to elite Southern schools: Frank Jr. went to Vanderbilt, Elliott and Robert went to Washington and Lee.[31]

However, like other builders, Frank Sr. frequently experienced both boom and bust. He went bankrupt in a recession following World War I, moved the family to Pennsylvania for a time, then reemerged in the mid-1920s as a leading builder in Birmingham, Alabama, where he constructed the city's Medical Arts Building and a twelve-story temple for the Shriners. Although back in the black, so to speak, and president of a Birmingham country club, Frank Sr. told the family "to be careful not to reveal [they] were Catholics in front of the Shriners."[32] When the Depression hit, he weathered the storm in Nashville, Tennessee, where he built homes in the city's Belle Meade suburb and also a bottling plant for Coca-Cola. Family legend has it that Frank Sr. was offered stock in Coca-Cola in lieu of cash, but, because times were tough, he unwisely chose the cash.

Nevertheless, by the late 1930s the Mackles were wealthy enough to vacation in Delray Beach, Florida, north of Fort Lauderdale, where in 1938 Frank Sr. purchased eleven city lots. He then asked his sons to "test" the Florida market by building houses there. As Frank Jr. remembered it, "Dad told the three of us . . . to go down and build homes on the lots. . . . We built real nice homes with tile bathrooms and oak floors, one bedroom, and sold [each of] them for $2,750." Then "we called and asked if we could buy eleven more lots" to build more homes, "and he said we could." When those sold, Dad said that "he'd better come join us."[33] Thus, in 1939, the Mackle Company returned to Florida, developing two small subdivisions, "Mackle Park" in West Palm Beach and "West Buena Vista Centre" in Miami, and a 200-home development for enlisted men at the Miami Naval Air Station at Opa-locka in 1940.

In 1941, Frank Sr. died at age sixty, but Elliott, Robert, and Frank Jr. continued the business by overseeing a massive $18 million expansion of a U.S. Navy base at Key West. (Robert and Frank Jr. would ultimately serve in the Navy—Frank Jr. as an officer in the "Seabees," the Navy's famed construction battalion—while Elliott took a deferment to run the company.) When the war ended, the Mackles returned to doing what it was they did best: building low-cost homes in subdivisions in and around Miami. "Don't build monuments to yourself," their father once told them. "Build what the public wants and what they can pay for."[34]

What they could pay for were houses in the $5,000 range—about $50,000 today—if they were subsidized by FHA or VA loans and sold with almost no money down. One Mackle ad from 1951 for a home in Miami's Tropical Park Estates featured a two-bedroom cinder-block house for $4,950, with $25 down and $25 due each week until a full payment of $250 was made, then a monthly payment of $29.44, which included taxes, insurance, principle, and interest.

The Mackles kept prices low, the saying went, by "buying by the acre and selling by the foot," and by using cost-efficient building techniques similar to those of Levitt & Sons in New York. Although not exactly prefabricated, the houses were built using studs and rafters delivered to the site in precut "packages." The Mackles measured windows and doors, indeed the entire house, in eight- and sixteen-inch multiples so cinder blocks wouldn't need cutting. Bathrooms abutted kitchens to keep drain pipes short, ceilings were unplastered, and the cabinets and closets had no doors.

The trick to a Mackle home, however, wasn't just in the design. It was in how it was built, in lightning-fast steps that looked less like a construction site and more like an automobile plant. "Lot preparation was done a block at a time," wrote one family member. "Layouts for houses were done with a wooden template. The labor crew dug the foundation. Carpenter crews formed the slab and laid the wire mesh. Concrete crews poured and finished the slab. Block masons erected the walls. And on it went. A crew—finished with one house—simply moved next door to do it all over again. And home construction was continuous for as far as the eye could see."[35]

The numbers were staggering: 450 houses at Grapeland Heights, 483 houses at Linden Gardens, 234 houses at Randall Park, plus 1,200 more homes, not to mention several commercial buildings, through April 1950 alone. These were in addition to the Mackle brothers' biggest project yet—turning a former coconut plantation off the coast of Miami into the middle-class island community of Key Biscayne. For years, the island had been accessible only by boat, but in 1947, the county raised funds for the 1.2-mile Rickenbacker Causeway, which opened the area for business. Sitting on 220 acres bought for $800,000 from the heirs of New York industrialist William J. Matheson, the Mackles' "Biscayne Key Estates" featured three-bedroom homes in the $11,000 to $15,000 range that included a club membership.

"They were cement block structures," wrote one chronicler, "sturdy" and "side-by-side." But "if they weren't designed for architectural awards, the setting among rows of coconuts and other palms" more than "made up for their plainness."[36] The island's luster was given a boost when the Mackles built the four-story Key Biscayne Hotel and Villas between 1951 and 1953. It was here that president-elect John F. Kennedy and then vice president Richard M. Nixon met for a postelection summit in 1960, and where Nixon vacationed before buying a residence at 500 Bay Lane, on the leeward side of the island, in 1969. "Nixon came and the world found us," said one early resident. "He and Pat and the two girls lived over one of the Villas at the Key Biscayne Hotel. . . . Pat would yell out the window to someone down at the Beach Club: 'Your phone's ringing!' We used to see them all over."[37]

Critics initially believed that no one would live on Key Biscayne because of the toll bridge, but the island proved so popular that by 1954 the Biscayne Key Estates, which had grown by 180 additional acres and added a yacht club, had sold out. (Today, the island's remaining Mackle homes go for as much as $1.8 million and are bought for the property, like the Nixon compound was, and torn down.) Meanwhile, and in addition to their Key Biscayne project, in 1954 the Mackles had big developments going in Pompano Beach Highlands in Broward County, where they planned 4,500 homes, and one in Westwood Lakes in Dade, where they planned 3,500 homes. As it turned out, the financing for Westwood Lakes came from a mining magnate named James H. Buchanan, president of the Florida Canada Corporation whose subsidiary, the Florida West Coast Land Company, had purchased the 78,000-acre Frizzell Ranch in 1954.

Buchanan and Frank Jr. were golfing buddies at the exclusive La Gorce Country Club in Miami Beach—the same club that would ban Michael Jordan in 2012 for wearing cargo shorts—where the two men struck a deal: Florida Canada would sell the Mackles some 6,200 acres north of Charlotte Harbor for $2.9 million. In exchange, the companies would form a fifty-fifty partnership in the community of "Port Charlotte" in which Florida Canada

An early aerial view of Port Charlotte looking south, circa 1957. Charlotte Harbor is in the background. Courtesy of the Randol and General Development Corporation (GDC) Collection, object ID CCHC-RGDCC-0003, Charlotte County (Florida) Libraries and History Division.

would finance roads, drainage systems, canals, and other infrastructure and provide additional land "as needed," while the Mackles would build homes. Their aspirations, at first, were small: a sales office on U.S. 41, a motel with a restaurant, four model homes, and 500 homesites, which they platted in September 1955.

Although Florida Canada had enough additional land for hundreds of thousands of lots, Buchanan promised officials that both he and the Mackles would "make haste slowly" in a "carefully planned program" that would be "a strong factor in the county's growth." He envisioned some light industry, even tourism, but said bluntly: "We will not be selling raw lots to widows and the aged at $200 apiece."[38] Actually, they were $895 apiece, and they were homesites instead of raw lots and sold at $10 down and $10 a month to retirees. "Now remember, my father and uncles were builders," insists Frank

Mackle III, who later headed the business. "They weren't into land sales. They developed land and improved lots to build houses. That's what they did. So the installment industry was something they came to by accident."[39]

It began in late 1955, when the Mackle Company published an ad in *Life* magazine for Pompano Beach Highlands, where homes started at $4,950. They received a whopping 18,000 inquiries, but only 127 people opted to buy. "We were sitting around, [the three of us]," remembered Frank Jr. in a 1992 interview, and "I said, 'We got to sell them something.' So we got the idea of ten dollars down on the lot only, the payments tied to retirements, figuring they would hire us to build the home when they could."[40] It was a revolutionary idea, last seen in Florida in the 1920s, but only a handful of developers, the Mackles among them, had thought to apply low-end land sales specifically to retirees. They tried it out through a limited ad campaign in 1956 and did a completely unexpected $2.6 million in sales, which was more than they'd made selling homes in Port Charlotte.

To capitalize, in 1957, the Florida Canada Corporation brought in cash for advertising by selling a 31 percent stake in the company to Canadian financier Lou Chesler, a one-time business associate of the gangster Meyer Lansky, who in October of that year announced a seven-week, $2 million media blitz in hopes of selling a platted section of 27,000 lots. Directed by the Mackles, the campaign started with a two-page spread in the *Saturday Evening Post*, followed by sizable ads in over 200 newspapers and magazines, including *Popular Mechanics*, *U.S. News & World Report*, *Time*, *Look*, *Life*, and *Reader's Digest*. The Mackles also made a promotional film on Port Charlotte, narrated by former NBC anchor John Cameron Swayze that aired on forty East Coast TV stations, and radio spots for an additional 175 stations.[41]

Unlike earlier Mackle ads—which sold houses—these new ads focused entirely on land sales and pitching 80′ × 125′ lots to the masses. "The old idea of Florida as a place where only the rich could live," exclaimed one promotion, "has been shattered by the vision and foresight of three brothers, Elliott, Robert, and Frank Mackle. . . . Imagine, being able to buy a piece of choice Florida land for only 'cigarette money!' A high, dry Florida homesite that can make your dreams of a future in Florida come true. . . . And it's so easy to do! Just fill out the coupon below and mail with a ten-dollar deposit. As soon as your deposit is received you will be sent a contract, and a map showing the exact location of your property. After your first deposit, you pay only pennies a day—ten dollars a month. . . . [So] act now! Send the coupon today. Remember, the future belongs to those who plan ahead. . . . Reserve your homesite now!"[42]

A 1957 newspaper ad for "The Mackle Plan." Retiring middle-class northerners would buy land sight unseen, receiving a map marked with an "X" where their lots were supposed to be. Courtesy of the author.

The response was astounding, and the Mackles' Miami offices were inundated with what one employee called a "bushel full" of cash. Another employee remembered Elliott Mackle yelling "I got a triple!" when he opened an envelope with not one but three separate coupons enclosed.[43] The buyer wanted three lots. Others wanted more expensive property on canals or street corners, while a small yet still sizable percentage wanted homes. In all, for 1957 the Mackle–Florida Canada partnership did $15.5 million in installment land sales, plus $6.7 million in home sales. It did so much business and processed so many coupons that it was forced to rent an extra office building for its new mail-order staff. As Frank Jr. put it: "We could sell 10,000 lots and 1,000 homes a month if we could produce that many."[44]

That was an exaggeration—the partnership had sold perhaps 40,000 homesites and 1,000 homes in Florida by late 1958—but the lesson was clear: installment land sales were the alpha and omega of Florida's boom. As one person noted, "installment land sales, a product which did not exist three years earlier, now made up seventy percent of the Mackles' business."[45] It grew even bigger, in April 1958, when Florida Canada and the Mackle Company merged their fifty-fifty partnership and three other partnerships into a publicly traded company called the General Development Corporation (GDC). The Mackles owned 13 percent, or 200,000 shares, receiving two and two-thirds percent of company land sales through 1962 and building contracts for all GDC homes. In exchange, the Mackle Company worked exclusively for GDC, where Frank Jr. was president and Robert and Elliott were vice presidents in charge of developing over 185,000 acres of land.[46]

In 1960, there were eight GDC communities, the largest of which was Port Charlotte, which, with additional land purchases, would balloon to over 100,000 acres. It was followed, most notably, by Port Malabar near Melbourne with 45,000 acres (encompassed today by Palm Bay, Florida), Port St. Lucie near Vero Beach with 36,500 acres, and Port St. John near Titusville with 5,400 acres. Of course, the prefix "Port" was a marketing ploy, for neither Port Charlotte nor any other GDC community had port facilities to speak of. In fact, the future Port LaBelle in rural Hendry County was forty-five miles from the ocean. They were all "eerily similar communities with nautical names," noted one journalist, and they followed a similar plan: thousands of rectangular lots laid out in a gridiron for as far as the eye could see. "Port Charlotte is one big chessboard," explains Charlotte County historian and land developer Robert "Bucky" McQueen. "I mean, they just cookie-cuttered it. I'm talking miles and miles of roads, way out into the boondocks. It made you wonder, what was more important to them, selling lots or building homes?"[47]

Once their lots were paid off, buyers would choose a home design from a limited number of low-cost models. In Mackle communities like Port Charlotte, these homes were so ubiquitous residents would say "I live in a Mackle." Courtesy of Charlotte County (Florida) Libraries and History Division.

The answer, at least according to GDC's own financial statements, was clear, because in 1959, the company made $56.7 million in land sales, which was five times more than the $10.6 million it made selling homes.[48] And, as the money grew, the marketing became more intense, for in addition to annual ad budgets in the millions of dollars, by the early 1960s GDC had sales representatives in at least 100 cities, plus branch offices in Florida and one in Germany that sold property to GIs. In 1959, the company began chartering bus, plane, and train trips to Florida and in 1960 it built a five-room model home on the seventh floor of the Mandel Brothers department store in Chicago. "We want the public to be able to make their own personal inspection of a typical General Development home," said Robert Mackle, "the kind they could live in at our Florida communities."[49]

Designed by in-house architects at the Mackle Company, the GDC homes in Port Charlotte came in ten pastel-colored models, which were the epitome of 1950s "Florida living." They were narrow, one- to three-room rectangles, with jalousie doors and windows, a screened patio, sliding-glass back doors, and a carport. Because air-conditioning was a luxury, the emphasis was on airflow, therefore the typical house was, at most, two rooms wide and the windows had louvered slats. Yards were minimally landscaped—GDC had bulldozed whatever pine trees or bushes were left before residents had gotten there—so the terrain was completely clear. "It was October 6, 1956, on my fourteenth birthday when we moved into that house," remembered former resident Connie Platt, whose family was the first to actually live in Port Charlotte. "There were no trees. It had been a cattle range, and the only landmark was a cabbage palm in the front yard."[50]

The Platts were from Chicago; four years later, in 1960, the Bicking family of Denver arrived with fourteen-year-old son Jody in tow. By then, Port Charlotte had grown to almost 3,200 residents, the same as Punta Gorda, and though it had a grocery store, a drugstore, a furniture store, a shoe store, a barber shop, and a 7-Eleven, it was, as Jody insists, "weirdly unfinished," with "a house here, then nothing; a house there, then nothing. All laid out in that big Mackle grid. That first trip, we came down [U.S.] 41 and took a left, and there was one house *waaaay* out in the distance. I mean way out there. And that was our house. Mom and Dad were trying to be upbeat, saying 'home sweet home' and all that, but my only thought was 'where the fuck are we?'"[51]

Certainly, as an unincorporated planned community, which in geographic terms was six times the size of Manhattan, Port Charlotte was odd. For one, it had no downtown, just a smattering of businesses housed in strip malls on either side of a highway—in today's parlance, a "commercial corridor." It also had, as of 1960, no city government because it wasn't a city; there was no industry apart from the building industry, no libraries, no hospitals, no sidewalks, and no schools. What it did have were affordable homes for retirees, whom the Mackles targeted in ad after ad after ad. There were thousands of them. In March 1960, for example, they ran (what had to have been) an incredibly expensive sixteen-page insert in both the *Chicago Tribune* and the *New York Times*, which asked senior citizens to "consider the most pressing need of this century . . . the need of people retiring on moderate incomes" to find "a place to live in healthful happiness" while protecting "against increases in the cost of living . . . the pensioner's nightmare."

With frequent references to long, cold northern winters full of "extreme discomforts" such as snow and flu, the Mackles described Port Charlotte as the antidote, the place that had "more days of pleasant outdoor life" than any other place, where the air was "clear" and "bracing" and where "mangrove-edged keys and islets" were "jewels of greenery that whisper[ed] of mystery and adventure."[52] It sounded exotic but was within the grasp of those who could shoulder $200 down and mortgage payments of $46 month. That wasn't a lot; the average retiree in 1960 earned over $80 a month in Social Security, so the Mackles planned Port Charlotte for retired couples earning twice that—the so-called medium-income retirees. In the past, people "played up Florida as a rich man's playground," said one Mackle exec. "Now we're finding that the mass market—of folks of moderate income—is far more important, and in the long run, far more lucrative for us."[53]

In 1960, the typical Port Charlotte resident was a retired tradesman: a printer, a mechanic, a carpenter, or a construction worker, maybe a fireman or a policeman. Some had been teachers, civil service workers, or factory workers, and virtually every one of them had come to Florida for the weather. "Supervising snow clearance was one of my biggest headaches, winter after winter," said a county engineer from New Jersey named Harry Skinner, "so you can imagine that when retirement approached, I was determined to move to a warm climate."[54] Widowed former plant manager Elmer Gray lived in a canal-front home with his daughter. He missed his friends in Ohio, he said, but told a reporter: "Do you know it's three below up there today?"[55]

Though polls showed "sunshine," "health," and "outdoor living" were the main reasons retired Americans moved to Florida, elderly transplants also favored its taxes—or, rather, its lack of taxes, for the state had no income tax and gave homeowners a homestead exemption of $5,000, with assessments set at just 50 to 75 percent of the home's actual value. Thus, in Port Charlotte, where the first Mackle homes cost between $5,975 and $7,650, many early residents paid no property taxes whatsoever. Others paid a trifling amount, considering that northern states had higher mill rates and some, such as New Jersey, had no homestead exemption at all. Yet even with these savings, to which one could add lower heating costs and fewer home maintenance costs, many retired residents still struggled to get by. A significant number lived Social Security check to Social Security check. Therefore, they opposed the imposition of local taxes vociferously while shooting down city incorporation efforts on numerous occasions and resisting even small increases for streetlights.

"What do we need them for anyway?" snapped J. C. Darden, a retired federal worker from Virginia in 1962. "If we go out at night we can always leave on our entrance lights. And as for prowlers, who'd want to prowl a community of old retirees with no romance, no money, no valuables, and nothing to offer but conversation?"[56] By a margin of more than two to one, Darden's neighbors agreed with him. They wanted extremely low taxes, a government limited to trash pickup, policing, and mosquito control, and to live their lives as cheaply and securely as possible. "Listen, I'm on a disability from the Army that gives me $119 a month and my wife and I get Social Security of $106 a month," explained a retired building superintendent. "In Long Island, I couldn't live on that. . . . But look at what I have here. If it wasn't for the Mackles, we wouldn't have half this."[57]

That desire to own their own homes, in Florida, during their retirement years—even without high incomes—was for northerners a "seemingly universal" dream.[58] In fact, in 1958, the Florida Development Commission reported that one out of every six Social Security recipients who left his or her home state left it for Florida.[59] The number-one destination for retired Ford workers, outside of Michigan, was Florida; the number-one destination for retired federal workers, outside of Washington, D.C., was Florida; and Florida had the highest ratio of retired military families to nonmilitary families, one to twenty-one, in the country.[60] They were enticed, of course, by cheap land, cheap homes, warm weather, and low taxes, but also by chambers of commerce stressing "friendliness" instead of "exclusiveness" and by developers such as the Mackles, who, as one resident put it, "were willing to build for the little guy."[61]

Nevertheless, it is important to note that these "little guys," these medium-income pensioner-transplants that Port Charlotte was built for, were consumers, and by the 1950s, American consumerism was intricately tied to the suburbs, to "white-flight," to the mass of Americans leaving cities to buy homes in outlying communities such as Park Forest, Illinois, or Mahwah, New Jersey, or Levittown, New York. As historian Lizabeth Cohen noted in her classic work *A Consumers' Republic: The Politics of Mass Consumption in Postwar America*, between 1947 and 1953, America's suburban population grew an astonishing 43 percent while the population at large grew just 11 percent. What is more, between 1950 and 1960, the country's twenty largest cities grew by just a tenth of a percent while their suburbs boomed by an equally astonishing 45 percent.[62]

But suburbs were for the young. Levitt & Sons geared its popular New York community and two other communities in Pennsylvania and New Jersey, for example, to young families, specifically white-collar families earning at least $5,500 a year. Blue-collar workers couldn't afford to live in the suburbs, explained Cohen, not even Ford workers at a glistening new automotive assembly plant in Mahwah. Although the factory had moved to the suburbs, the workers continued to live in the city.[63] Thus, retired northerners, many of whom earned far less than Mahwah's workers did, faced a choice: they could stay in increasingly decrepit, racially divided, highly taxed urban areas, or they could buy homes in suburbs of their own—in this case, in low-end retirement communities in Florida.

"[We] retired folks on fixed incomes came to Port Charlotte," explained a New York pensioner in 1962, "to get away from the cultures, crimes, graft, [and] corruption" of northern cities and also their "taxes."[64] But, like suburban residents in Mahwah or Levittown, they too wanted homes, which the Mackles and other Florida developers were happy to provide. The result was a wave of migration and retired folks of all economic stripes coming to Florida. Between 1950 and 1960, the state's age 65-and-over demographic exploded by almost 250,000 people to 553,129. In Port Charlotte, whose 3,157 residents had appeared seemingly overnight, 668, or 21.2 percent, were age 65 and over, a number twice the national average. Fully 36.5 percent, or 1,152 people, were age 55 and over, while septuagenarians and octogenarians outnumbered teenagers by more than two to one. Curiously, older men outnumbered older women, though men tended to die earlier; married residents outnumbered single residents; and because of discriminatory FHA policies favoring racially segregated neighborhoods, not to mention GDC's own sales practices, there were no African Americans in Port Charlotte.

"Many of the older people who first settled" in Port Charlotte "call themselves 'pioneers,'" wrote *Life* magazine in 1959, and they "submerge their pangs of homesickness in the feeling of being adventurous."[65] There were, for example, various "Pioneer" clubs, including one called the "Pioneer Handicrafters Club," which met each week at a house on Martin Street, where the community's first model home was. "[We had] bulls in the front yard and bulldozers in the back in those days," said one GDC salesman, so residents golfed, fished, played cards, went to the beach, or held boat motorcades, known as "boat-a-cades," to stay busy.[66] They were keenly aware of which residents were from where—Port Charlotte's semiannual directory listed not only names and addresses but also places of origin, so Claude and May Chronister of 381 Waltham Street were from Harrisburg, Pennsylvania; William and Margaret Pike of 228 Olean Boulevard were from Watertown, Massachusetts; and so on.

Fearing homesick residents would leave Port Charlotte without organized activities to keep them entertained, in 1961, the Mackles opened a cultural center at the corner of Tamiami Trail and Easy Street called "Port Charlotte U." It had to have had "the oldest student body in history," quipped one journalist, and the unaccredited curriculum included courses in typing, stenography, Romance languages, lawn-mowing, and marine navigation.[67]

The registration fee was two dollars. "People got bored after [GDC] brought them to Port Charlotte," explained an administrator at today's Cultural Center of Charlotte County, a successor to Port Charlotte U. "They said, 'Hey, you brought us here to this lovely place and there's nothing to do!'"[68] There was nothing to do because in the early 1960s, there were no enclosed shopping malls anywhere in Florida, at least not in the modern sense, and no pro sports teams (the Dolphins came in 1966, the Bucs in 1976), no Interstate 75 or Interstate 95, and no Disney World. The closest town of any size was Fort Myers, whose 22,500 inhabitants were thirty miles away. Tampa was 100 miles away, while Miami was 185 miles to the south and east through a Miccosukee Indian reservation in the Everglades.

To be sure, Port Charlotte had been modestly successful—with over 2,000 homes, a core of community buildings and businesses, and 3,200 residents in four years—but it was far smaller than the city of 70,000 that GDC said would be there by 1965. In fact, critics wondered if that particular prediction had been a lie, if Port Charlotte itself was a "Potemkin Village" of sorts intended to show northerners just enough amenities to trick them into buying lots. "For all their pious pretenses," wrote journalist Anthony Wolff in his 1973 book *Unreal Estate: The Lowdown on Land Hustling*, these "planned communities . . . are neither planned, nor communities. They are lot sales operations, pure and simple." Wolff named Port Charlotte in particular, where "settlers," he wrote, "may take full advantage of the facilities the promoters have provided for window dressing. But most of the subdivision will probably lie fallow forever."[69]

In 1974, in a candid interview with INFORM, an environmental organization studying the impact of installment land sales in America, GDC's Port Charlotte project manager, Dougald Ferguson, admitted that the community had "started out as a land-sales project. Then core community development, initiated to generate new sales, [had] snowballed the site into a city." But, because planning had been so poor, said Ferguson, and houses built randomly, "it's too bad we can't go back." The problem was that, in building Port Charlotte, neither GDC nor the Mackles had practiced what experts call "phased development," in which lots were platted, sold, and built upon in consecutive sections. Instead, wrote INFORM, "virtually all lots were platted in the subdivision's early years, and were sold as fast as the company could market them."[70] As a result, Port Charlotte (and a second, neighboring development known as North Port Charlotte, today's "North Port") had too many lots—a reported 198,104 by 1976—leaving much of the community a ghost town.

"The whole place is subdivided into little neighborhoods of the future," wrote one observer, "and crisscrossed with more streets than Tampa and St. Petersburg have between them. That's enough blacktop to pave a two-lane road to Seattle. And more than 90 percent of those roads go nowhere."[71] There were so many roads, in fact, that GDC's public relations director Tom Ferris had trouble naming them all. "Easy Street" was a no-brainer; so were streets named for Mackle family members, such as "Barbara Avenue," or Mackle employees, such as "Ferris Drive." Unbeknownst to most residents, there was even an "Elkcam Boulevard," a name common to nearly all Mackle developments, which was Mackle spelled backward. Ferris then went through every New York–related place name he could think of, apparently, before having an idea: He would name streets for the real estate editors of northern newspapers in a shrewd if not craven attempt to garner publicity. As a result, today's Port Charlotte features a "Herron Terrace," named for *Washington Post* editor Paul Herron, a "Hemenway Avenue" for *Hartford Times* editor Allyn Hemenway, and so forth. And, since Ferris began his career at the *Toledo Blade*, Port Charlotte has a "Toledo Blade Boulevard."[72]

Unfortunately, most of the streets named by Ferris remained empty for decades, becoming trash dumps or make-out spots or impromptu landing strips for Cessnas loaded with drugs that flew in from the Caribbean.[73] Indeed, some streets laid out more than sixty years ago in the northwest part of Port Charlotte have been completely abandoned. Cracked and covered in grass, they have a creepy, postapocalyptic feel reminiscent of Chernobyl. Reporter Gary Roberts visited the area in 2015. "The rustic terrain of northwest Port Charlotte," he wrote, "is, at the same time, beautiful and uninviting. A red-shouldered hawk swoops down . . . gliding effortlessly over untamed wilderness. Meanwhile, a few blocks away . . . a swarm of turkey vultures devour the remains of a ten-foot alligator." As Roberts reported, there were 17,412 residential lots in the area—all GDC creations—of which some 15,000 were vacant.[74]

This twenty-first-century backlog of 1950s homesites is a direct consequence of what GDC did. It divided land prematurely, before anyone anywhere in the world would live there, to sell at $10 down and $10 a month. There's actually a term for it: "premature subdivision," and by the mid-1960s, thanks to GDC and other corporate developers, Florida had over a million prematurely subdivided lots.[75] It seems counterintuitive, perhaps even wasteful, because the most basic improvements, such as land grading or ditch digging, cost money. But for GDC, the payoff wasn't in Florida, per se. It was on Wall Street, where an accounting practice known as "accrual ac-

The be-all and end-all of 1950s Florida land development was the 80′ × 125′ lot. By the early 1970s, Port Charlotte and its sister community North Port would have nearly two hundred thousand 80′ × 125′ lots, each connected by a road. Courtesy of the Charlotte Harbor Area Historical Society and Ulysses Samuel (U.S.) Cleveland Collection, object ID FX-11A, Charlotte County (Florida) Libraries and History Division.

counting" allowed the company to claim the entire price of a lot, upfront, as revenue in the year it was sold. So even though a $895 lot brought in, at most, $120 a year in twelve $10 installments, GDC counted the entire $895 as received income. It was, as one observer put it, "an obscure detail . . . overlooked by lay investors," but a detail that GDC manipulated "to goose the goose that laid the golden eggs."[76]

Starting at about $10 a share in April 1958, when Florida Canada merged with the Mackles to create GDC, the company's stock price would explode nearly eightfold to a high of $77.50 in 1959. Its sales figures were astronomical: $38.4 million in 1958, $67.2 million in 1959, and $56.1 million in 1960.

Although certainly inflated by land sales, GDC's revenues did include, by 1960, over $15 million in home sales, which the Mackles had asked other GDC board members and chairman Lou Chesler to emphasize. "The development and sale of land," explained Frank Mackle Jr. in a 1964 *Miami Herald* interview, "profitable though we have proved it to be, is not an end in itself. It is only the prelude to our real objective: the construction of full communities."[77]

Mackle wasn't being altruistic. Land sales were shady, as brokers who'd never been to Port Charlotte, let alone other GDC communities, frequently used high-pressured sales tactics in which they overstated the value of lots and sometimes lied. Not everyone did this, but what had started as a limited ad campaign in 1956 had morphed by the 1960s into a billion-dollar, installment land sales industry. From Florida to California, GDC competitors were carving up enormous tracts of land, selling lots in planned communities or peddling unimproved acreage in uninhabitable places like the Mojave Desert or the Everglades.

When a number of these developments proved fraudulent, among them the infamous "Lake Valley Estates" in eastern Oregon, which promised buyers they could fish and swim, presumably in a lake, when the closest body of water was thirty-five miles away, the entire land sales industry, and with it GDC and Mackles, became tainted.[78] In fact, by the early 1960s, the National Association of Better Business Bureaus had received a tidal wave of complaints, leading president Victor H. Nyborg to declare that "the Florida land promotions of the 1920s pale in significance . . . to what is happening now. When this thing breaks, and it is going to break, thousands and thousands of people are going to raise hell."[79]

For the Mackles, overly aggressive sales techniques weren't just unsavory or even illegal, they threatened to turn the company into a lot-sales-only operation whose entire *raison d'être* was to cold-call pensioners from a boiler room, which was not their original intent. But high-pressured sales pitches would make GDC a fortune selling lots. The more it sold, the higher its (inflated) income; the higher its (inflated) income, the more valuable its stock, and no one owned more GDC stock than chairman Lou Chesler, who at one point had over 1.4 million shares. Chesler favored land sales; in fact, he and GDC board members Max Orovitz and Gardner Cowles wished to expand them while the Mackles did not. They wanted home sales—a shift in company focus—but when their contract expired in 1962, the one that had given them the exclusive right to build GDC homes at a commission, Chesler forced them out. They sold their shares for a reported $5.85 million in February 1962 and in December formed their own land giant, the Deltona Corporation.[80]

CHAPTER TWO

# The Land Giants

> I am summarizing the evolution of six or seven years now, when what started with a magazine ad developed into a fleet of airplanes, a printing plant, a room full of WATS-line jockeys, and salesmen everywhere from San Francisco to the Vatican.
>
> —JOHN ROTHCHILD, *Up for Grabs*

A tall, dignified man in his forties, Cornelius McGillicuddy Jr. was known for two things. First, as vice president of Florida's Gulf American Corporation, a subdividing leviathan that oversaw the sprawling Cape Coral development near Fort Myers, he was a fixer, a doer, a straight shooter, and a problem solver who managed crisis after crisis that owners Jack and Leonard Rosen had caused. Second, Cornelius McGillicuddy Jr. was the son and namesake of Cornelius McGillicuddy Sr., the famed Philadelphia Athletics baseball coach known to sports fans as Connie Mack. Both Macks, junior and senior, had come to Fort Myers in the 1920s, when the team trained each spring at the city's Terry Park. In 1951 Connie Mack Jr. came to stay. He sold real estate, worked as a banker, and owned a shrimping boat before joining the Rosens in 1958.

If Gulf American had a face, he was the face, and everyone who visited Cape Coral or saw the company's promotional materials knew his name. After all, the senior Mack had coached the A's for over fifty years, winning nine pennants and five world championships, while his calm and upright demeanor, not to mention his frequent pronouncements against gambling and alcohol, made him the most trusted man in baseball. When he died in 1956 at age 93, President Dwight D. Eisenhower called him an "inspiration," while none other than Ty Cobb, the mean-as-dirt brawler and batting champion who once played for Mack, cried.[1] For lot sellers like the Rosens, therefore, hiring the son of such an esteemed personage, who was a local no less, made sense.

"It will never be known how many people bought Cape Coral homesites solely because of meeting Connie Mack, Jr.," remembered Gulf American employee Eileen Bernard, or were influenced by "knowing he was part of the company, or getting a letter signed by him. His signature and name were used in thousands of ways. He officiated at every event, big or small, always friendly, his dignity enhanced by his six-foot-five-inch frame."[2] The junior

The Mackles' success with installment land sales led to the creation of more than 200 competing companies, including the Gulf American Corporation, founded by former mail-order shampoo salesmen Jack (left) and Leonard (right) Rosen. Courtesy of the Cape Coral Historical Museum, Cape Coral, Florida.

Mack, with his aura of respectability, stood in marked contrast to the Rosens, who had few if any ethical boundaries to prevent them from selling land, especially Leonard, who had a magnetic, in-your-face quality that was impossible to resist. He was also a manipulator.

In his 1993 book *Selling the Dream: The Gulf American Corporation and the Building of Cape Coral, Florida*, historian David E. Dodrill interviewed an early Rosen assistant named Bernice Freiberg, who told the story of how Leonard Rosen once sold coat hangers at a county fair. "Leonard would pitch the hangers and [have an employee] say, 'Can I have two?' and Leonard would say 'I'm sorry, only one to a customer.'" To which the employee would plead, "'But I really need two. One for my brother-in-law.' And Leonard would say, 'You better get out of here. I'm going to call the police. Only one to a customer.'" In that way, he stoked the crowd, explained Freiberg, and soon, "all these people [were yelling], 'give me two, give me three.'"[3] That was in the 1930s. By 1940, the Rosens had sold enough cheap items and other bric-a-brac to open a furniture and appliance business that peddled goods in installments—"Cash or Twenty-Four Months to Pay," read one ad—and while Leonard traveled the Eastern seaboard training salesmen, Jack handled the books.

They sold out in 1952 to concentrate on their newest venture: the wholesale marketing of Charles Antell haircare products and a shampoo called "Formula No. 9." The latter was nothing more than bulk shampoo mixed with lanolin, the wax from sheep wool, which the Rosens blended in a bathtub then sold on TV. Using terms like "lustrous," "shimmering," "replenishing," and "restorative," they extolled lanolin as a wonder ingredient capable of curing dandruff and preventing hair loss. As one commercial asked: "Have you ever seen a bald sheep?" The answer was no, but lanolin had nothing to do with it, and neither did Formula No. 9. Talk show legend Ed McMahon, who himself began as a carnival barker and knew the Rosens personally, once likened the brothers to snake oil salesmen and joked, "Believe me, there were no formulas 1 through 8. . . . People had been selling tonics out of the backs of wagons for a hundred years," and to the Rosens, television was "the new covered wagon."[4]

The brothers used the new media intensively, sponsoring game shows like "What's Your Bid?" and cutting deals with TV stations to run what, in effect, were the first infomercials. They also peppered magazines like *Life* with some truly audacious ads, including one in 1952 depicting a frazzled woman with a noose around her neck holding a gun in one hand and poison in the other, who exclaimed, "If my hair looks such a mess *one more* night, I'll kill myself!"[5] Revolting as they were, the Rosen's Charles Antell ads were incredibly effective, so much so that by the mid-1950s, Formula No. 9 was one of the most recognizable shampoo brands in the United States. Other Charles Antell products, such as diet pills, children's vitamins, and makeup, were popular as well. But, as Charles Antell expanded, the Rosens found it increasingly difficult, not to mention expensive, to gain or even maintain market share while competing with industry heavyweights such as Revlon, so in 1956 they started looking for their next big thing.[6]

The idea came to them in Miami, where Leonard Rosen went to recuperate from an arthritic condition and where, out of pure serendipity, the former appliance salesman turned blitz-marketing guru witnessed the first stirrings of the Florida land boom. On his return to Baltimore, he and his brother studied a short film on Lehigh Acres, a Port Charlotte–like development to the east of Fort Myers, then sent Bernice Freiberg to Florida for a clandestine tour and sales pitch in Indian River County at a small General Development Corporation (GDC) community in Vero Beach.[7] "I went in there pretending to be a customer," remembered Freiberg in a 1988 interview.

"They gave me a purple piece of paper with prices on it and maybe a little four-page handout," but that was it—nothing the Rosens would consider salesmanship. Emboldened, the brothers then bought an eight-acre parcel in Deland, Florida, near Orlando, which they subdivided and test-marketed through a thirty-minute infomercial in Baltimore.

The response was overwhelming, so in late 1956 Leonard Rosen returned to Florida to look for property. He was in Punta Gorda, seeking treatment for arthritis at the old Punta Gorda Hotel, now a spa, when he ran into Milt Mendelsohn, a former ad executive turned land developer from Chicago.[8] Known for his intelligence, his lack of scruples, and his meticulously artistic disposition, Mendelsohn had first worked in installment land sales in Lehigh Acres, but in 1955 formed the Charlotte County Land and Title Co., which bought and subdivided a 2,000-acre spread to the east of Port Charlotte called Harbour Heights. The land had belonged to a retired pilot and brigadier general named Charles H. Danforth, who once had a home there, but it was swampy in parts and expensive to develop.

Mendelsohn dredged what he could, which wasn't much, built two model homes and some roads, then marketed Harbour Heights as "a fashionable Florida subdivision." He then distributed a slick, full-color brochure detailing a "country club type recreation center with matchless facilities," a private set of docks with anchorage, tennis courts, a golf course, a pool, and also a tarpon tournament. As an added bonus, or as Mendelsohn called it, a "gracious extra," the company would plant, without charge, "a blooming, ready to bear orange tree on your homesite when you [were] ready to build."[9] Yet more than a year in, Mendelsohn had built or provided none of those things, but had sold an ungodly 1,700 lots, some of which were underwater.[10] "My guess is that while Leonard was [visiting Punta Gorda]," commented Kenny Schwartz, the future Gulf American manager of Cape Coral, "Milt was telling him about some of the opportunities" to be had down there, and that "caught Leonard's imagination."[11]

Even so, the elder Rosen knew Mendelsohn was a crook. He liked him and decided to hire him—first, to scout out potential properties, then to plan their development—but as he later told an employee: "Milt Mendelsohn never signs anything. . . . Listen to what he says and learn from him" but "he's never allowed to make any commitment on our behalf."[12] It was a good thing, too, because Mendelsohn's Harbour Heights development became a bankrupt ghost town, while he later pled guilty and served eighteen months—at age seventy—for bilking investors in a second Florida development called "Rocket City."[13] But, whatever the case, in early 1957, Leonard

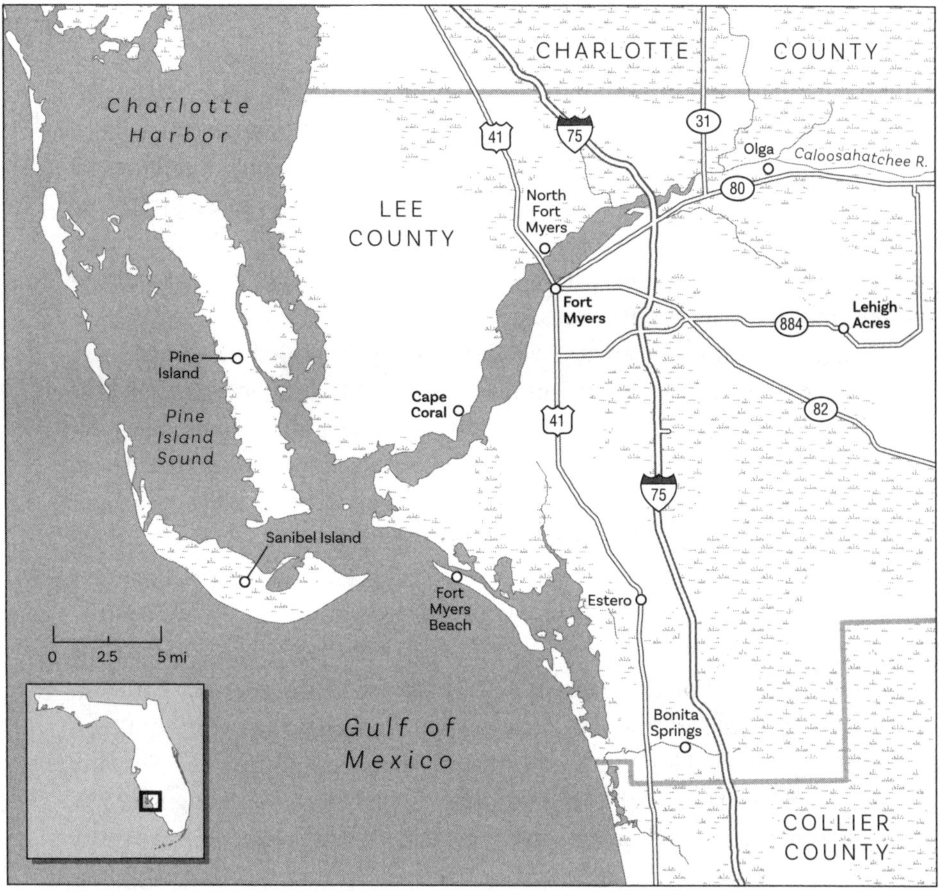


Lee County, Florida. Map created by Erin Greb Cartography.

Rosen and Milt Mendelsohn began touring Southwest Florida by airplane, and Rosen chose a 1,700-acre tract thirty minutes south of Punta Gorda across the Caloosahatchee River from Fort Myers. Like Punta Gorda, Fort Myers was originally a frontier city founded in the 1800s. It drew fishermen, cattleman, and some tourists—including inventor Thomas Edison, who built a laboratory and spent winters there beginning in 1886—and it was surrounded by hydric pine flatwoods, prairie land, and swamp.

Even by local standards, the land Rosen wanted, called "Redfish Point," had been laid waste by tree cutting and turpentining. The land was marshy, and grass for scrub cows and other animals in its upland sections was poor. Homesteaders called it "Hungryland," and virtually everyone who lived there, with the exception, perhaps, of a former preacher named "Wild Bill" Belvin, struggled to survive. Dubbed the "Lee County Robinson Crusoe," in 1930, at the behest of a publicity-seeking news editor from Fort Myers, the forty-eight-year-old Belvin attempted to prove the area was inhabitable by spending a year in the bush, living "without the aid of civilization's conveniences."[14] Belvin was dropped off in the heart of Hungryland with only a bathing suit on, which he quickly shed, and could be seen periodically setting traps and harpooning stingrays and wandering about in an outfit made of moss. He exited the woods in good health in 1931 but was jailed almost immediately for hunting pelican eggs out of season.[15]

As for the Rosens, their 1,700-acre parcel belonged to the heirs of Dr. Franklin Miles, a former Fort Myers resident and Indiana native whose Miles Laboratories sold both One-A-Day vitamins and Alka-Seltzer. The price was $678,000, for otherwise worthless land, of which the Rosens contributed just $125,000. They had investors, mostly friends and business associates from Baltimore who helped form (what became) the Gulf American Corporation, and though they sold Charles Antell in 1958 for a cool $2 million, they spent as little of their own money as possible.

With Mendelsohn in charge of planning, the Rosens chose the name Cape Coral, which was odd, because the Caloosahatchee had no coral, but Mendelsohn liked the alliteration. (He would give similar names to other Rosen developments in Florida called Golden Gate, River Ranch, and Remuda Ranch, and one in Arizona called Rio Rico.) The Rosens' lead engineer was Tom Weber, who came to the project in late 1957 after a career in the Army building radar installations in such inhospitable places as Iceland and Brazil. It was a good thing, too, because unlike Port Charlotte, where much of the land had been drained through canal building in the early 1900s, the original Cape property was incredibly low, made up, to a large extent, of environmentally sensitive wetlands covered in mangroves, salt marshes, and sloughs.

Therefore, to raise the land to at least 5½ feet above sea level, which was the minimum needed for building, Weber turned to an old Florida technique: dredge and fill. First practiced in the 1920s by a West Virginia native and Fort Lauderdale land developer named Charles Green Rodes, dredge and fill meant digging a canal, then using the mud and other sediment to make land, usually finger-like peninsulas with rows of houses on either side bordered by a seawall.[16] This was the epitome of Florida living, in which each house had a driveway in the front and a boat dock with ocean access in the back. Although devastating environmentally—canals destroyed habitats and put salt in the aquifer, for example—"finger islanding," as this practice was

In 1958, the Rosens began developing Cape Coral, a planned community near Fort Myers. Using dredge and fill techniques to raise the land above sea level, the Rosens' Gulf American Corporation and its successor companies would build upwards of 100,000 homesites with more canals—some 400 miles of them—than any other city in the world. Courtesy of the Cape Coral Historical Museum, Cape Coral, Florida.

called, was a win-win for the Rosens. First, it created land, dry land suitable for homesites, where previously there was none. But more importantly, it turned this new land into waterfront property, a valuable commodity for which buyers paid premium prices. "If you've got waterfront property," wrote one journalist, "you've got what the rest of the population is looking for." So "developers buy up barren land, dredge out black-water trenches leading to canals or bayous miles away, and sell the land at enormous profits."[17]

Indeed, the Rosens had bought Redfish Point for $393 an acre; however, within a year, they sold canal-front properties for $1,990 a lot and river-front properties for $3,390. When, in 1960, they purchased the adjacent 16,000-acre Matlacha Ranch from investment banker Ogden Phipps, their plans for Cape Coral, Florida's "waterfront wonderland" they called it, went into overdrive. With four huge dredgers, each a flat-bottomed barge equipped with a rotating digger in the front and powerful pumps to suck up sediment and shoot it, onshore, in the back, the Rosens had dug 168 miles of canals by 1963. In the coming years, they and Gulf American's successor corporation, GAC Properties, dredged even more. As a result, today, Cape Coral boasts more canals, an exorbitant 413 miles of them, than any other city in the world, including Venice, Italy. Unfortunately, Cape Coral's original canals were sometimes twenty-five-feet deep, too deep to maintain necessary oxygen levels for coastal marine life, while the dredged fill dirt was sandy, full of limestone, and poor.

In fact, as fill dirt rose, it tended to suffocate whatever scrub pines or other vegetation was left, so rather than landscape creatively—for example, by building wells around larger trees or planting new trees—the Rosens bulldozed the site into piles and let it burn. What remained was a flat, bare prairie, a treeless moonscape with the steppe-like ambiance of Kazakhstan. Cape Coral's "environmental impact has been, in a word, disastrous," stated one 1977 report, and "Gulf America's environmental planning was non-existent. The project was started in an era when wetlands were regarded as useless swamps until touched by a developer's magic wand."[18] Even so, the Rosens themselves were unapologetic, even glib. As brother Leonard once said: "If, by raping the land, they mean taking land that was absolutely useless and raising it above sea level, then that's what we did. And if they mean we got the highest prices for the land that we could, then, yes, we did that too."[19]

STARTING IN JANUARY 1958 with a six-person sales office but no houses, the Rosens would do over $9 million in business by the end of the year. They relied at first, as the Mackles did, on mail-order land sales, but as the industry grew, they turned to independent and company-run brokerages in cities throughout the United States and even opened offices abroad. Gone was the coupon—by 1960, only 2 percent of GDC sales came through the mail, for example—and in its place was the sales pitch, over the phone from a call center, or face-to-face at info booths in train and bus terminals, at dinners or cocktail parties hosted by the company, or on tours. Soon, there were 234 major developments in Florida—with the state defining "major" as any development offering fifty or more lots through installment plans in a given year—and they all used a similar sell, the hard sell, to bring in hundreds of millions a year.[20]

By 1965, Gulf American had surpassed GDC, as well the Mackles' Deltona Corporation and the Lehigh Acres Development Corporation, to become the largest of the land giants with over $100 million a year in sales. Its rise was

Entrance to Cape Coral, Florida, late 1950s. The Rosens had no experience with land development when they founded Cape Coral, but their incredibly aggressive, sometimes comical sales practices made the Gulf American Corporation the largest land developer in Florida. Courtesy of the Cape Coral Historical Museum, Cape Coral, Florida.

nothing short of remarkable, but the industry itself was remarkable, perhaps even surreal: "Hey baby!" said a beautiful, winking Thai girl to a passing GI at the entrance to a Bangkok bar. "I've got something to show you!" As the military newspaper *Stars and Stripes* reported, he "didn't require a second invitation." The soldier "eagerly squired the girl . . . to a large room where they found other girls and other GIs laughing, talking, and drinking." Then, "after everyone was suitably settled down, the real pitch came. They saw movies of Florida property belonging to the company who had hired the girls and financed the booze."[21]

It was a practice known as "bird-dogging," in which civilian girls, Army support personnel, and sometimes other GIs were paid by corporations to solicit military buyers. Companies even had offices near bases. The GDC headquarters in Frankfurt, Germany, for example, was directly across from the giant I. G. Farben Building, home of the U.S. Army's V Corps, from whence retired soldiers, now GDC reps, ferried customers back and forth in Volkswagen buses with the words "Port Charlotte," "Sebastian Highlands," and "Port St. Lucie" emblazoned on the side.[22] Retired soldiers knew the lingo, as did General Bruce Easley, who worked for the Gulf American Corporation, and General Bruce C. Clarke, hero of the Battle of the Bulge, who handled military relations for the Horizon Corporation and who authored a flyer entitled "Why the Military Man Should Acquire Land."

The land giants knew why, because apart from northern retirees, military men were the perfect audience, representing, just at European installations, some $30 million a year in business. "They're travelled and bright," said one insider, but "they've been rootless for years and look eagerly to the day when they can settle down." Plus, "color brochures and 'easy payments' often cause them to develop a 'what can we lose?' attitude."[23] Corporate "bird dogs" also targeted American tourists abroad, who paused their sightseeing trips in Rome or London to listen to a sales pitch. According to Joe Raso, a sales manager at Gulf American's Rome office in the 1960s, the company used "good looking young girls" to lure Yankees into a lunch at the Hilton and a film on Cape Coral. "We had sales people that could hardly speak English telling these American people . . . how great Florida was and what a good investment it was, and people would buy. My first month there, I had a million dollars' worth of sales."[24]

Not to be outdone, foreigners bought, too, because by 1965, GDC was selling more than $4 million a year to Europeans and Latin Americans; Deltona, which at its height had forty-six overseas offices, did $1 million a year in Hong Kong alone.[25] The sister and mother of Philippine President Ferdinand Marcos bought property in Cape Coral, as did an unnamed "Latin gentleman dressed in a resplendent uniform" who took $17,000 out of a money belt but "didn't make a dent in it," wrote Eileen Bernard.[26] Nevertheless, the land giants' real bread and butter was the States. Here, thousands of independent agents and brokers, company-run salesmen, WATS-line callers, and "bird dogs" of every stripe, including bikini-clad girls at swimming pools at tourist hotels in Miami, all played bit parts in a real-life reenactment of the stage play *Glengarry Glen Ross*. Author and Tampa-native Paul Reyes wrote as much in his 2010 work *Exiles in Eden: Life among the Ruins of Florida's Great Recession*, in which he recounted how in 1969 his parents agreed to a daylong bus trip from Miami Beach to desolate Lehigh Acres after being cold-called . . . in their hotel room . . . on their honeymoon!

Established by Chicago businessman and "d-CON" rat poison founder Lee Ratner, who had made millions as a mail-order marketing guru, Lehigh Acres was a 100-square-mile former cattle ranch located on scrub and timber land

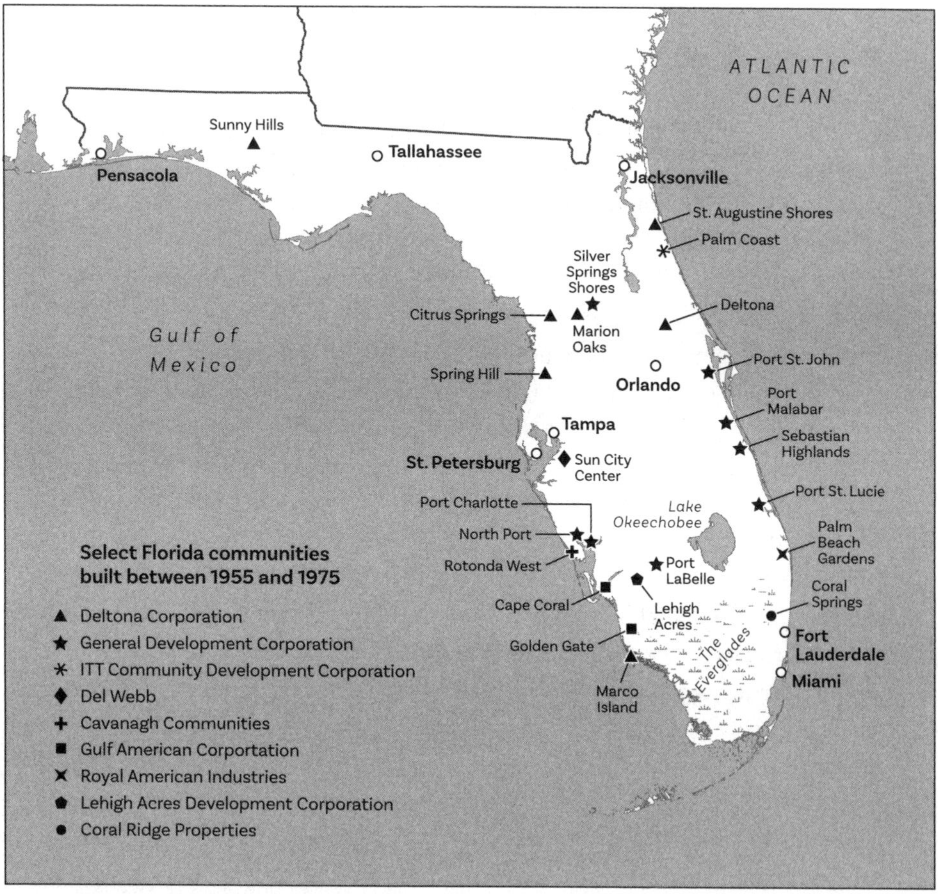


By the mid-1960s, the success of the installment land sales industry in Florida led to the creation of some 234 companies offering fifty lots or more through installment plans. The largest of these companies were known as "land giants." Listed on Wall Street, they owned hundreds of thousands of acres and built entire communities from scratch. Map created by Erin Greb Cartography.

some seventeen miles east of Fort Myers. Although Lehigh Acres was exceptionally remote, not to mention landlocked, in 1955, Ratner and former advertising executive Gerald Gould began marketing the community as "the area that has everything—everything that millionaires and the international set pay fabulous sums to enjoy." That was untrue, for all Lehigh Acres had, apart from pine trees and palmetto bushes, was a particularly nasty species of wild hog. However, Ratner and Gould went even further, printing a twenty-page brochure with the insane caption "Thomas Edison discovered it!" What they meant (but didn't say) was that Edison had "discovered" Fort Myers, not Lehigh Acres, so they coyly included an artist's rendering of a palm-lined thoroughfare, a golf course by the water, and even a beach.[27]

The whole thing, admitted Gould in a 2009 interview, was a scam. It wasn't a community, he said, at least not in the beginning. It was "a marketing concept." "We gave so much thought to selling the land that the normal reservations for commercial properties, schools, all the ancillary things you need in a community, weren't made. We even had canals that ran uphill."[28] By 1969, the year Reyes's parents arrived, the Lehigh Acres Development Corporation had sold some 80,000 homesites but had built just 3,000 homes. The couple came by tour bus, in which an excited driver described in glowing terms what they were traveling through: Drainage canals! Sugar cane fields! Alligators! It was almost as if Florida's swamps and prairie land were . . . prairie land and swamps. But, writes Reyes, "My parents were sufficiently tenderized by the time they got to Lehigh Acres. . . . Sore and unsuspecting [they and other] couples were led through a dazzling tour of model homes, then corralled into eight-by-eight offices to be worked over by hucksters peddling quarter-acre lots with the greasy hubris of a Bible salesman."[29]

The couple bought a lot, then a week later, the bride's parents did, too. It was a process repeated hundreds of thousands of times in shady developments all over Florida. In Cape Coral, where buses from Miami were called "Sucker Buses," Gulf American gave free airplane tours in tiny company Cessnas and used a street as a runway. By 1961, the company's "Fly and Buy" program had planes taking off every eight minutes. Flying five days a week, six hours a day, they carried over a hundred thousand passengers a year.[30] These local flights were in addition to chartered airline trips, called "sponsored trips," in which northern sales representatives sent buyers to Florida in carefully coordinated tours. Devised by the Mackles in 1959, the sponsored trip was a four-day junket, advertised in the real estate sections of newspapers, which included room, board, and transportation for as little as $50. "Go A.W.U.T.F.," declared one 1968 ad by the General Development Corporation. "Go Away with Us to Florida! Just $50 round trip!"[31]

The only problem was that some "sponsored trips" by many of the more disreputable operators were gimmicks, the classic case being the "Win-a-Week-in-Florida" contest held at supermarkets or fairs in which nearly everyone won, but no land sales company was mentioned and little was free. The rooms were free, provided "winners" paid a $15 "overhead and processing fee" and attended what one contest called a "Fabulous Florida Dinner-Show," which, in effect, was a sales pitch. That contest "has more strings than the

New York Philharmonic," quipped the head of the Rochester Better Business Bureau, because "not only does the winner have to pay his own travel expenses, he has to spend about one-fourth of his time listening to a [salesman]. It's required, and if you skip it, you may find yourself billed for 'free' accommodations."[32] Sometimes, northerners would "win" a contest, travel to Miami or Fort Lauderdale, then hear for the first time that a tour of a land development was mandatory. As one Ohio man asked, incredulously, "Why was the sale of land in Lehigh Acres never mentioned [to us] until after we arrived in Florida?"[33]

Yet even with bad publicity, by the late 1960s, "sponsored trips" brought in some 60,000 people a month.[34] That's 720,000 people a year, more than the combined populations, in 1970, of Rochester, Dayton, and Providence traveling to Florida to visit land developments and listen to sales pitches. Now add those to thirteen different Miami-based phone centers, where each year an estimated 600 "regional directors" made some 4.5 million calls. As journalist Elizabeth Whitney wrote in her 1970 series "The Swamp Peddlers," the person "on the other end of the conversation no doubt has conjured up a vision of a high-powered executive, immaculate in his business suit and tie." But "that immaculate high-powered executive" is actually "a cross between a middle-aged hippie and a pool-hall habitué."[35] Of course, callers weren't hired for their appearance; they were hired, first and foremost, for their ability to lie.

In fact, in 1971, Whitney published a secret call-center script from a company named Royal Palm Beach Colony, Inc., the developer of Royal Palm Beach Village on the east coast of Florida, which included no fewer than six easily verifiable falsehoods, from the old "buy now or the special sales price I told you about will expire on such-and-such a morning," to claims that Palm Beach County was the fastest-growing county in the state. (It was eighth.) And, to assure potential customers that Royal Palm Beach Village was going to be built on dry land, the caller made the claim that, in topographical terms, Florida was shaped like a spoon, "higher on the edges than in the center and this is why you have the population on the two coast lines today."[36] (The interior of the state, in general, is 100 to 200 feet higher than either coast, which is right at sea level.)

Indeed, the moral vacuity of land company call centers knew no bounds, for none other than Connie Mack Jr. was shocked to discover, sometime in the 1960s, that every one of the hundred or so Gulf American employees at its call center in Miami, many with foreign accents, would begin conversations with: "I'm Connie Mack Jr."[37] Although Mack flew into a rage and confronted the Rosens personally, putting a stop to the practice, the bottom line was that for Gulf American and other land giants, the cold call was pivotally important, because even with a seemingly low 5 percent success rate, which is what Whitney estimated, 4.5 million calls annually brought in hundreds of millions of dollars in sales.

So did a practice known as the sales dinner. "Please be our guests for a special reception and dinner, Tuesday, April 20, 1965 at the Holiday Inn North," read an invitation from a Gulf American broker to a couple in Daytona Beach. "You are assigned to Table No. 17. Informal dress." The couple wasn't interested but gave the invitation to Morton Paulson, the business editor of the *Daytona Beach News-Journal*, who, skeptical of land giants, attended the dinner himself, incognito, and recorded his impressions in print.[38] (The dinner was meant to sell Gulf American homesites in a new development near Naples called "Golden Gate Estates.") "All of the twenty-nine guests," wrote Paulson, ages forty-five and up, "gave the appearance of solid, cautious, well-off citizens who weren't about to be conned into anything." That was before the dishes were cleared and two silver-tongued salesmen took the mike.

"Is anybody here allergic to making money?" asked William G. Swilley, a drawly Florida man with the bearing of a preacher. The audience members, who sat in small groups at seven or eight tables, each with a sales "counselor," shook their heads. "Now, let me make this clear. We're not going to sell you anything tonight. Couldn't if we wanted to. All we can do now is accept your reservation." By "reservation," Swilley meant "down payment," which, he explained, could be refunded anytime in the next six months, provided the person buying the property visited Golden Gate.

At this point, Swilley was interrupted by a salesman named Gerald Earney, a short man with a piercing voice. (Earney, noted Paulson, had previously worked at "Rocket City," the failed Milt Mendelsohn development near Orlando.) He opened with a pep talk, extolling the virtues of Gulf American and telling the audience that what the company had done in Cape Coral it would do again in Golden Gate.

Along the way, he skillfully addressed (or downplayed) the audience's most obvious concerns. For example, as to whether or not people should buy land sight unseen, Earney asked: "When you buy a can marked spinach, how do you know what's in it?" To those short of cash he said: "Do you know 92 percent of Americans live beyond their means?" He then showed a film on Golden Gate, a slick production with quotes from Andrew Carnegie and Teddy Roosevelt, among others, plus scenes of cranes and earthmovers at work. "As you can see, they're not knee deep in water or fighting off alligators," opined Earney, though in fact, Gulf American had carved parts of Golden Gate Estates out of a swamp.

As the film finished and the lights came on, the counselors went for the kill. Paulson's counselor was Swilley. "You could really benefit from this," he intoned. "We'd like to get you started on an investment program tonight." The down payment was $105 on a $2,120 homesite, with a 2 percent bonus discount, and $100 refund for visiting.

"We'd like to think about it," said Paulson, who'd come with his wife. "How about letting us have a brochure and copy of the contract?"

"Now what can you possibly know in a day or so that you don't know right now," replied Swilley. "You can change your mind within six months—SIX MONTHS—and get every cent back. Where could you get a better offer than that?"

"Well, I just don't want to rush into things," countered Paulson. "I'd be signing up for over $2,000 in property and . . ."

"Do you realize you'll make $140 immediately when you sign this paper?" Swinney exclaimed. "There's the 2 percent bonus discount and the $100 deduction for going down there."

"Fine!" said Paulson. "Just let us have the brochure and the contract. I'd like my attorney to look at it."

"You sign that contract now," answered Swinney, "and show it to your attorney tomorrow and if he finds a single thing wrong with it, I guarantee you we'll tear up that contract and back comes your money."

At this point, Paulson's wife intervened. But "we just paid our income tax and don't have $105 in the bank," she said.

To which Swinney retorted: "We always hear that sort of thing from people. Do you realize that 92 percent of people in America live beyond their incomes?" The Paulsons got up to leave, but a flummoxed Swilley blocked the door.

"What seems to be the problem?" asked a counselor named Frazier.

"Oh, nothing much," replied Swilley. "They don't seem to have the cash."

"Is THAT all?" exclaimed Frazier. "Did you tell them that 92 percent of Americans live beyond their incomes?"

Swilley had, of course, but the Paulsons didn't bite. They pushed through the Swilley-Frazier phalanx, but just as they exited, Frazier called out: "Remember, procrastination is the mother of disappointment!"

The Paulsons were lucky. On occasion, sales counselors insulted reluctant buyers, such as in the fall of 1970, when Gulf American's successor corporation GAC Properties hosted a dinner in New Orleans that was attended by a *St. Petersburg Times* staff writer. "We were going 'round and 'round," remembered the writer, until "finally, [the salesman] got up and said to us, real disgusted, 'Let's get these losers out of here. There are some winners coming in soon.'"[39] Truth be told, if anyone was a winner at these events it was the land giant, because, as the *Wall Street Journal* reported in 1966, big companies had a sales-to-participant ratio of one-to-ten. (The Mackles' Deltona Corporation did even better, it reported, at one-to-five.) At four dollars a head, with lots by that point in the $2,000 range, the dinners were a phenomenal proposition, albeit for the company. "It's the best vehicle to display our product to the greatest number of people in the fastest possible time at the lowest practical cost," said one executive, who admitted that 40 percent of his invitation list came straight from the phonebook.[40]

Although dinner parties, cold calls, out-of-state brokerages, and international offices represented the lion's share of land company sales, customers who signed contracts in Florida, on-site, were less likely to ask for a refund, which is why Gulf American, GDC, Deltona, and the Lehigh Acres Development Corporation did whatever they could do to lure tourists to their communities.[41] In North Port, for example, GDC built the National Police Museum and Hall of Fame, where visitors paid an admittance fee to be fingerprinted and admire such items as the tear-gas riot gun. In Cape Coral, Gulf American planted a 40,000-bush rose garden (which subsequently died), installed a giant German-made fountain with choreographed lights and music called "Dancing Waters," and hosted a porpoise show. Deltona introduced the much ballyhooed "World Chicken Plucking Championships" in Spring Hill, where women had the shapeliness of their legs judged in the "Miss Drumstick Contest," while Lehigh Acres maintained a Polynesian Garden and a "state-of-the-art" hydroponic farm with an observation tower.

Naturally, the land giants had celebrity spokesmen on staff—the Mackles, for example, employed former Yankee first-baseman Bill "Moose" Skowron as a sales representative in Spring Hill and beamed host Don McNeil's long-running "Breakfast Club" radio program direct from Deltona—but each paid thousands more to bring even bigger celebrities to their events. In 1962, Liberace played the Lehigh Acres Auditorium, while in 1966, Bob Hope received the first (and only) "Patriot-of-the-Year" award from Jack and Leonard Rosen in a public ceremony in Cape Coral. In addition, developers would subsidize Hollywood films and TV shows if companies would bring the productions on-site and hold pageants, boat races, tennis and golf championships, and chess, bridge, and fishing tournaments too numerous to mention.

Even poor Harbour Heights got into the act. In 1961, some three years after founder Milt Mendelsohn had left for Cape Coral, it finally got around to hosting a tarpon tournament. (As a publicity stunt, organizers used a helicopter to pick up and transport a boat some twenty-three miles to the mouth of Charlotte Harbor at Boca Grande Pass.) The promotion didn't work, apparently, because shortly after its second tournament in 1962, the company went bankrupt.[42]

However, taken together, tourist traps and gaudy promotional activities were one thing; paying agents to harass visitors at major Florida theme parks was another. Although most companies did it, as usual Gulf American took things to the extreme, especially in Tampa, where the company purchased a 200-car parking lot across the street from Busch Gardens and put up a billboard that read: "Free Parking! See River Ranch Acres at Our Expense!" (A raw-acreage subdivision in eastern Polk County, River Ranch was one of the worst land scams in the state.) "Of all the complaints [we've received] since Busch Gardens opened," said park spokesman Robert Taliaferro, "we've had more on this parking lot operation . . . than everything else combined."[43] While the Rosens called their solicitors "Outside Public Consultants," or OPCs, company executive Joe Miller had a different name for them: "hustlers." In fact, in a 1988 interview, Miller claimed that Gulf American had opened its various Cape Coral attractions and had even purchased a Miami alligator farm called "Musa Isle" because its "hustlers" at other parks kept "being arrested."[44]

But when your boss was Leonard Rosen—a man who once told his salesmen that people "are here to be taken. If you don't take them, someone else will. [And] if you feel you don't want to take them, leave now"—your entire sales operation was a free-for-all.[45] It seemed, in fact, that by the early 1960s the entire industry was a free-for-all, one whose mantra was SELL, SELL, SELL! Indeed, employees at the Rotonda West development near El Jobean, founded in 1970 by the land giant Cavanagh Communities, had a story they liked to tell. One day a salesman was driving an elderly man and his wife across the Everglades from Miami, which in those days was a long, meandering trip, when the man had a heart attack and died. There were no cell phones or call boxes then, so as the salesman drove the body to the hospital, he used the time to convince the grieving widow to buy a lot.[46]

Although clearly apocryphal, the Rotonda story wasn't that far from the truth, for as Gulf American assistant Eileen Bernard wrote in her 1983 memoir *Lies That Came True: Tall Tales and Hard Sales in Cape Coral, Fla.*, a significant portion of salesmen would say anything and do anything to get YOU to

A Gulf American salesman closes a deal in Cape Coral, circa 1959. Led by Leonard Rosen, the company's sales force engaged in a variety of deceptive sales practices and reportedly spied on customers by placing listening devices in their closing rooms. Courtesy of the Cape Coral Historical Museum, Cape Coral, Florida.

sign YOUR name on the dotted line. "By 1966," she wrote, "most 'average' salesmen had been weeded out—or gave up in disgust or despair—and only the hard-core professional remained. No wonder customers were caught like lambs in the bramble bush."[47] The white lies were outrageous and frequently hilarious. While out showing property, for instance, a salesman would point at the largest house in sight and say, "That's my house." Or, if a woman was working outside, he'd say, "That's my maid." Or, if there was a boat, he'd say, "That's my boat." Good-looking salesmen were paired with single women customers while salesmen with religious backgrounds handled preachers and priests.

Once, according to Bernard, a salesman who'd gone to seminary told a group of churchgoers how sorry he was they were leaving because he was giving a sermon the next day. He wasn't, but when the churchgoers changed their travel plans specifically to see him, he lied a second time, telling them, "Oh, I'm sorry! I can't make it. The company needs to me to work"—which only caused the churchgoers to wonder if they should be buying from a company that did business on Sundays. When, on occasion, a customer demanded to see his land but there were no roads, a salesman would simply show him one that was accessible, then lie and say, "This is your property." As Bernard explained, a salesman named Dick was fired when buyers randomly showed pictures to each other of lots they'd purchased featuring the *same* tree. But he was fired, she wrote, "not because of his sins. . . . [He just] wasn't a good salesman."[48]

Both Gulf American and GDC eavesdropped on clients. In Cape Coral and Golden Gate, for example, Gulf American installed listening devices, actual bugs, in its closing rooms to hear what couples were saying. And, in Port Charlotte, GDC owned the Ramada Inn, where staff screened phone calls to and from clients' rooms to make sure outside real estate agents weren't able to speak to them. The company also removed newspapers from the lobby—because newspapers had ads from competing developments in them—and tried their utmost to keep customers from even seeing other options. "I was directed to make certain that I pick up GDC customers as early as 8:00 A.M.," said a former sales agent. "I was to wine and dine them all day . . . and in general wear them out to make certain they saw no other builders or realtors and saw no homes other than GDC homes."[49]

Unsurprisingly, salesmen for most land companies dissuaded African Americans from buying property, either by ignoring black neighborhoods in their cold calls or sales dinners or by giving those who came to their developments a less-than-enticing tour. "Let's put it this way," said former Gulf American executive Leonard Rich, our salesmen "didn't encourage them." They were "very aware that people who were going to retire to Florida, as a whole, being American, would be prejudiced against blacks," so they "simply didn't encourage them."[50] One salesmen at Punta Gorda Isles, a waterfront development dredged from swampy wetlands to the southwest of Punta Gorda in 1958, was a former Army officer who recruited officers himself. He traveled to bases mostly, and during his very first trip, at his first-ever sales presentation, he encountered an African-American group of Army doctors and their wives. They were legitimate buyers, but when they came to Florida for a tour, one of the development's bosses told the salesman, pointedly: "They really don't want to live here. Do you understand what I'm saying? They *really* don't want to live here."[51]

Although racially restrictive housing covenants had been ruled unenforceable with the Supreme Court's Shelley v. Kraemer case in 1948, Federal Housing Administration (FHA) loans and various other types of loans continued to be denied to African Americans until 1968 with the passage of the Fair Housing Act. And, while land companies had removed racist wording from their radio ads and print materials by at least the 1950s, salesmen across the board continued to discriminate informally. As a result, by 1970, Cape Coral had 10,193 residents, but just one African American. Lehigh Acres had 4,394 residents but just three African Americans, while Port Charlotte, with 10,769 residents, had none.[52]

AS FOR THE MACKLES, the brothers broke with General Development in 1962 to form the Deltona Corporation, so-named because its first planned community was on a lushly wooded, 15,000-acre tract in east-central Florida close to the cities of Deland and Daytona, hence the word Deltona. They purchased the property on the sly, in pieces, in hopes of preventing owners from holding out or jacking up prices, and they cobbled together one of the best inland locations in the state. Just thirty-five minutes north of Orlando and forty-five minutes west of the coast, Deltona had more than two dozen lakes, access to spring water from the great Floridan aquifer, and exits on Interstate 4, the cross-Florida expressway connecting Tampa to Daytona Beach. Lots were $10 down and $15 a month for seventy-seven months, with homes ranging from the $6,960 one-bedroom "Norwood" to the $15,450 three-bedroom "Oakwood." And buyers paying in cash got a 5 percent discount.

It was Port Charlotte, part two, except this time the Mackles made several badly needed logistical improvements to what GDC had been doing. First, they limited the community, as well as all future Deltona communities, to 15,000 acres; it was still a lot of land but far smaller than the 100,000 or so acres (in North Port and Port Charlotte) that GDC had been selling. As the Mackles had learned, in the field of installment land sales, bigger was not always better, especially when thousands of people who bought lots from you decided to put those lots on the market. GDC wouldn't buy them back—nor would any of the land giants—so owners listed property themselves or used outside real estate agents to sell it for what it was worth. That's why GDC, Gulf American, Lehigh Acres, and other land giants absolutely hated real estate agents, not to mention classified ads, because their prices, instead of being set by companies arbitrarily, represented a more realistic measure of what the market would bear.

The result was that developers sold lots at one price, usually to northerners and foreigners at sales dinners, while agents living locally sold lots for another price. "Meet Sharon White, first grader . . . real estate tycoon!" read one 1965 ad from the Gulf American Corporation. "How can a first grader be a real estate tycoon? It's easy. When Sharon was a baby, her daddy bought a Cape Coral homesite for her. He believed that as Sharon grew, so would the value of her property. Now here's the best part: Sharon's daddy paid

$1,590 for her homesite," and "today, similar property at Cape Coral is priced by Gulf American at $4,800." But was it worth $4,800? No. Gulf American claimed it was, but real estate ads in the Fort Myers *News-Press* that year were full of Cape properties selling at or below cost. The market was flooded with them. "Cape Coral lots sold at 1960 price," read one 1965 ad. "Buy equity and assume twenty-two dollars a month." Another read: "Builder's delight, package deal, 7 lots, 80′ × 125′, $1400 each."[53]

In Port Charlotte, an aggrieved former-customer-turned-real-estate-agent named Bernie Santaniello started his own business in the 1970s, looking up the names of absentee landowners using tax records, then offering them $1,000 a lot. Most had paid "at least that, sometimes twice that" through years of payments and interest, wrote one journalist, and though some were offended by the offer, others were happy to unload. Deed in hand, Santaniello would then flip the property for $1,700 a lot. But GDC sold comparable properties for upwards of $4,000 a lot, even telling the *Tampa Tribune-Times* in one 1981 article that a corner lot bought by Santaniello from a Pennsylvania widow for just $1,000 was somehow worth $10,000. Here's how GDC explained it: "On our international market, that lot—a corner lot—could well bring $10,000 because worldwide, there is a shortage of good Florida investment property. Locally, where all these lots may be viewed as a surplus, the value might appear to be somewhat less. . . ."[54]

Might appear? The value was less because anyone who paid $10,000 for the property had clearly been hoodwinked. That, said Santaniello, is what led him to start the business. He'd bought a lot in Port Charlotte from a salesman in Massachusetts, he explained, but "when I came down and saw it, I said, 'Oh, my God! They'll never build on this thing in my lifetime!'" He continued: "Look, we got 290,000 lots down here. The most that ever got built on in a year is 1,000. At that rate, it'll take 290 years to get built out there. . . . So I gave it to 'em. . . . Sure, [people I send letters to often] criticize me. But, I'd rather get that than hear from people who pay four times too much for a GDC lot, [and who are] sick and tired of making payments for twenty-seven years then coming down here to see what it's worth."[55]

To thwart Santaniello, who recorded 280 different land deals in 1980 alone, and other "independents" like him, companies would file complaints with the Florida Real Estate Commission and local realty boards and sometimes sue. In 1962, for example, Gulf American took legal action against Granville "Pete" Petrie, a former company pilot and salesman who had the audacity of becoming the first independent broker in Cape Coral and who represented a rival development nearby called "Coral Point." Gulf American alleged that the name Coral Point too closely resembled Cape Coral and that its employment of Petrie was "intended to confuse and mislead the public."[56]

The company would change its name, but when Petrie kept selling there (and also in Cape Coral), Leonard Rosen made several attempts to force the broker out. First, he had a giant hedge planted on Petrie's property line so no one could see his sign. Then, he had trucks parked in his driveway to dissuade customers and had a county attorney (who had also been a Gulf American attorney) falsely claim that Petrie's marquee violated the county's fifty-foot setback rule. Then, as Petrie's salesmen handed out flyers on the sidewalk, he had them sprayed with insecticide.[57] Obviously, Rosen was a brawler, but even he should have known that in planned Florida communities like Cape Coral, where by 1965 there were 71,627 landowners, resale markets were inevitable. The Mackles did, which is why they limited Deltona to just 33,000 lots and sold almost all of them.[58]

The Mackle brothers' second innovation, and something they hadn't done in Port Charlotte, was to implement what they called "coordinated growth." In stage one, they'd fill out the central core of Deltona with houses and lots, then move outward in four additional stages with lots paid in cash, lots paid in two- and three-year installments, and so on. This allowed the company to develop Deltona intelligently, at least by standards in 1960, and to avoid building a scattered community with houses way out in the woods, because those were expensive, said the mayor of Port St. Lucie, an early GDC development, in 1991. When "people [are] buying and building all over . . . you [can't] provide services at an economic cost. You know, garbage, police, roads. The basic infrastructure [here] could have been a lot easier if it had grown from the center."[59]

And finally, the Mackles' third innovation was to provide central water and sewerage systems to Deltona's core homesites—that is, to homes built in "phase one" of the development from its inception. Although unthinkable today, when Port Charlotte, Cape Coral, Lehigh Acres, and other such communities were built, state and local regulations allowed developers to simply forgo sewer systems in favor of septic tanks. They also didn't require builders to hook up homes to water lines, either; so instead, virtually all early residents used wells. In Port Charlotte, for example, GDC sold homes beginning in 1956 but wouldn't build its residents a water treatment plant until 1960. Even then, lines extended only to certain neighborhoods; therefore, many people used well water and continue to do so to this day. In fact, according to a report issued by the South Florida Water Management District in 2015, Lehigh Acres had just over 110,000 residents that year, but an eye-popping 30,000 wells.[60]

With so many wells drawing from one aquifer, the region's water table has dropped, massively, and water scarcity and water quality issues have plagued developments like Lehigh Acres for decades. And, since neither state nor federal agencies inspected private wells for contaminates—and still don't—residents there were out of luck. They sometimes paid an installment fee, had to do testing and maintenance, and the water they pumped had a high sulfur content that made it smell like rotten eggs. Even worse, in some developments, where soils were porous or sandy and wells reached just twenty feet down, heavy rainfall would overwhelm septic tanks to push sewage effluent directly into the water table. When that happened, explained a Fort Myers health official in 1966, residents, in effect, were "recirculating their own sewage." Bad enough at one house, he continued, but "the problem grows acute when a well-settled subdivision has dozens of [water] pumps and dozens of septic tanks arranged in alternate rows."[61] They were supposed to be seventy-five feet apart, but even at that distance, one 1970 study found that 19 percent of wells in northwest Melbourne, Florida, contained harmful bacteria, while in Brevard County, home of the GDC development Port St. John, some 30,000 septic tanks released more than 10 million gallons of waste water into the soil every day.[62]

Thus, for the Mackles, having water and sewer systems at Deltona was a selling point, one with the added bonus of providing the company with a second revenue stream: utilities. They founded "General Development Utilities" in 1959, a GDC subsidiary that by 1962 owned ten water plants and ten sewage treatment plants and served over 20,000 people.[63] So, rather than build utilities after the fact, which is what GDC, Gulf American, and Lehigh Acres had done, the Mackles installed Deltona's infrastructure from the get-go. They "clustered" development in phase one, then provided water and sewer through a subsidiary called "Deltona Utilities." It made sense: sell land, build houses, then, when the lots ran out, become a utility company. However, water and sewer lines were expensive, so developers who built them retroactively attempted to hook up customers as quickly as possible, with some, such as the Lehigh Acres Development Corporation, even persuading people to trade lots in the hinterland for those with water and sewer lines closer to the community core. As company president Gerald Gould put it, "We tried to do the right thing. Maybe it was a little late. Maybe it was a lame effort. But we tried."[64]

Yet, lest the Mackles receive too much credit, it should be noted that Deltona's sewer system was never really expanded and that apart from homes built in phase one of the development, everyone else used tanks. They still do: As of 2015, only 15 percent of homes in Deltona were connected to sewer lines, which meant 26,000 homes there were (and continue to be) off the grid.[65] Flooding was another problem, and though the Mackles designed a stormwater system that was, as one expert put it, "ahead of its time," it, too, wasn't finished. In 1969, for example, the community's Elkcam Boulevard flooded, and the company used boats to ferry residents back and forth across the roadway to their homes.[66] People also accused the company of unfair business practices, such as in 1971, when its franchised New York agent Florida Properties Unlimited sued Deltona for making "false claims" about its homesites, or in 1972, when a retired Chicago bus driver testified before an Illinois Attorney General's committee that when he traveled to Spring Hill, Florida, to inspect a lot he had purchased, he wasn't allowed to see it.[67]

However, overall, as Deltona prospered and the Mackles broke ground on three additional communities at Marco Island, Citrus Springs, and Spring Hill, and even entered the mobile home business, they retained a reputation for honest dealing. Environmental issues, infrastructure issues, and high-pressure sales tactics were simply part of the business. In fact, in 1964, none other than Robert Caro, a *Newsday* correspondent who famously excoriated Gulf American in print, testified before a Senate subcommittee that the Mackles "had the financial ability and the sense of moral responsibility not only to promote a retirement city, but to develop it over [the] long term."[68] One official called them "the squares of the industry," because unlike other developers, they eschewed street solicitors in places like Rome or Hong Kong and limited lot sales to foreigners to just 15 percent of the business. Typically, foreigners bought lots for investment purposes, explained a Deltona representative, so "if everybody buys [as an] investment, the investment is no good."[69]

Devout Catholics and avowed family men, the Mackle brothers were exceptionally close. They shared an office and a secretary at a company headquarters in the Miami neighborhood of Coral Way, close to downtown Miami and the Rickenbacker Causeway to Key Biscayne. "We're probably the only corporation in the U.S. with three presidents," said Frank Jr. "We have three desks in one big room. Robert handles the accounting and signs all the checks, Elliott oversees the construction end and the acquisition of property, and I handle the corporate financing and stockholder relations. We were brought up to work together and for a common end."[70] That "common end" was making money through a seemingly endless stream of land deals and businesses, not only in large communities like Deltona, Spring Hill, and Marco Island, but in a dozen or so smaller ventures all over the state. These ventures included, among others, "Sunny Hills," a planned community halfway between Panama City and Chipley, Florida, in the Panhandle, where the Mackles had purchased 25,000 acres; Lake Maitland Terrace, an extensive condominium complex near Orlando; plus a hotel on Marco, a lumber company in Bartow, and a horse ranch near Ocala called "Elkcam Farm."

The Mackles were racing fanatics and bred and trained thoroughbreds together for almost thirty years. As the story goes, in 1948, the brothers bought themselves two yearlings as a Christmas present and, because it was Christmas, they raced the horses under the colors red and green. However, their mother didn't approve. She'd never liked racing and had once told the boys, "Only Vanderbilts and bums fool with racehorses."[71] So, rather than use the Mackle name, out of deference to their mother, they called the operation "Elkcam Stables" at "Elkcam Farm." They raced between six and twelve thoroughbreds a year, all in red and green, and entered not one but two horses into the 1959 Kentucky Derby. The horses placed eighth and ninth, but the brothers enjoyed a memorable trip from Miami to Louisville with their families in a private Pullman car.[72]

In terms of volunteer work, the brothers served in leadership positions or on the boards of various nonprofit organizations, such as the Florida Catholic Children's Home, the Catholic Welfare Bureau of Miami, Mercy Hospital, St. Francis Hospital, Barry College, Biscayne College, and the University of Notre Dame, where Frank Jr. was a lay trustee. Politically, the brothers were conservatives, supporting presidential candidate and frequent Key Biscayne visitor Richard Nixon in his 1968 campaign, as well as Florida Governor Claude Kirk, who in 1967 named Elliott Mackle to the five-member Florida Land Sales Board. The board had existed in one form or another since 1963, when Florida officials realized, belatedly, that existing laws and state oversight mechanisms had failed miserably in regulating not only the land giants but hundreds of other developers, some of whom used false advertising and fraudulent sales tactics to peddle lots. It is here, on the Florida Land Sales Board, that Elliott Mackle, and by extension the Deltona Corporation, would go head-to-head in a knockdown, drag-out dispute with the Rosens.

CHAPTER THREE

# Into the Cattle Chute

> The parcels at River Ranch not only had no roads but also had no electricity, no drinking water and no sewage system. . . . But that did not seem to discourage potential buyers, few of whom ever saw the actual piece of property they were buying and some of whom were never sent, or overlooked, the fine print in the public offering statement.
>
> —DENNIS COVINGTON, *Redneck Riviera*

In 1967, Florida's recently elected governor was a gregarious, free-wheeling Republican named Claude Kirk. Tall with a Southern drawl, jowly good looks, and a love of jiujitsu, in 1966 the forty-year-old former insurance executive shocked observers by defeating Miami mayor and Democratic candidate Robert King High to become Florida's first Republican governor since Reconstruction. As one newspaper quipped, "The last one [we had] was a one-armed carpetbagger who was turned out of office ninety years ago," so having a Republican governor "is going to be a novelty to us."[1] Truth be told, Kirk was a novelty to everyone. A truth-bending, attention-seeking, self-aggrandizing populist, Kirk was Donald Trump before there was a Trump, wrote longtime Florida political reporter Martin Dyckman. "He didn't even know how many members there were in the legislature," remembered Florida House Republican leader Don Reed. "He didn't even know what Cabinet officers there were, he didn't know anything about state government. And he didn't bother to learn a whole hell of a lot about it while he was governor. He spent a lot of time enjoying being governor."[2]

Kirk was always good with a quote, telling *Time* magazine in one 1967 interview that he was "a tree-shakin' son of a bitch."[3] He made headlines when, at his inaugural ball, the divorced father of two brought as his escort a beautiful blonde woman whom he introduced to reporters only as "Madame X" (in fact, she was a German-born Brazilian named Erika Mattfeld and soon to become his wife).[4] Kirk's platform was full of contradictions. He promised to cut taxes, though the budget he signed contained the largest tax increase in Florida's history. He vowed to defy a federal judge who ordered the desegregation of schools in Bradenton, but, as one journalist put it, "skedaddled back to Tallahassee" when faced with a fine.[5] He promised to fight cor-

Elected in 1966, the grandstanding Florida governor Claude Kirk disliked the Rosens personally. Therefore, he strengthened the state regulatory agency that oversaw installment land sales and forced the brothers out of business. Courtesy of the Florida Photographic Collection, State Archives of Florida.

ruption, even hiring the private Wackenhut Detective Agency to ferret out wrongdoers, but maintained a pay-to-play slush fund called the "Governor's Club" while charging the Florida Development Commission for both his wedding pictures and his honeymoon.[6] (He later called it "a clerical error.")

The governor was all over the place—literally and figuratively. He once descended 1,000 feet below sea level in a submarine eight miles off the coast of Miami where he planted flags of Florida and the United States as way of drawing attention to what he called "the commercial possibilities of the sea." Ever the showman, he radioed reporters from the ocean floor, where he deadpanned: "I've just located President Johnson's popularity rating. It's 955 feet below sea level."[7] Unhappy with Florida's 250-mile-per-hour state-owned airplanes, he leased a 575-mile-per-hour Learjet, which he used for trips to New York or South America for his own businesses and sometimes took to New Orleans for lunch (and he charged the state and the Republican Party for them, until the latter cut him off).[8] Kirk's confrontations were legendary. In West Palm Beach he told the president of the famed "Outlaws" motorcycle gang that if he hadn't been governor, he would have fought him; he snatched a microphone from the hands of Black Panther H. Rap Brown to tell an African-American crowd in Jacksonville that "troublemakers weren't welcome in Florida," traveled to and berated officials in Satellite Beach for closing down an eleven-year-old's Popsicle stand, and took on Leonard Rosen.[9]

Rosen had been a Robert King High supporter and had not only backed the Democrat financially but had used Gulf American's headquarters in Miami as a campaign call center. Kirk didn't like Rosen. In late 1966, when the governor set up a postelection slush fund to pay for his campaign debts and aides asked wealthy businessmen from both parties to contribute to it, Rosen reportedly pledged $25,000, but only if he could give the money to the governor himself, in cash.[10] This was a no-go: Donations to a slush fund were one thing; being handed money in a face-to-face meeting with Florida's most controversial land developer was another. Of course, Rosen knew better. He most likely was goading Kirk and had little respect for either the governor or for the state agency he appointed, the Florida Installment Land Sales Board (FILSB). Established in 1963 in the wake of bad national publicity following thousands of consumer complaints to the Florida Real Estate Commission, the Better Business Bureau, local chambers of commerce, and state attorneys general throughout the United States, the FILSB was meant to protect buyers from unscrupulous land salesmen.

With jurisdiction over developers selling fifty or more lots annually through installment plans, the FILSB required companies, wherever they were headquartered, to register with the state, to submit all promotional materials for review and approval prepublication, and to give buyers a "property report," which was a basic description of the land being offered, including its location, topographical characteristics, the presence of utilities and drainage systems, and any easements, restrictions, or limitations affecting the land's occupancy. The FILSB also had subpoena power and could suspend a developer's license and those of its salesmen, while individuals who violated its regulations faced fines of up to $5,000 and six months in jail. The FILSB seemed tough, and though it would shut down several unlicensed or deceitful operators, developers who peddled less than fifty lots (so-called forty-niners) or lots through deed and mortgage weren't under its purview.

Another limitation was that in 1963, when Democratic Governor C. Farris Bryant and his legislative majority created the board, they hamstrung the agency from the beginning. As pro-business, pro-development, laissez-faire politicians, they hated regulation, so the goal in creating the FILSB was "to weed out unscrupulous land developers" while leaving the larger industry intact.[11] They had to do something, otherwise federal agencies such as the Securities and Exchange Commission or the Department of Housing and Urban Development, not to mention several influential senators from northern states whose constituents had been burned by land frauds, would. "There's a thing called the Florida Installment Land Sales Board," wrote one observer. "It's a phony. It was spawned to give a cloak of respectability to gluttonous land sales practices," but "it's a gimmick, pure and simple, to keep federal heat out of the state."[12] As proof, critics cited the sheer minuteness of the board—it had, for example, just eleven employees, including a bookkeeper and a janitor—plus three of its five members, by statute, were individuals "directly engaged in the sale or development of real estate subdivisions." In other words, they were insiders. In fact, the very first land sales board included Elliott Mackle and Lehigh Acres president Gerald Gould, which prompted journalist Morton Paulson to quip that "the foxes were guarding the chickens."[13]

Clearly, the relationship between land developers and politicians in Florida was a close one that led to absurd instances in which governors appointed developers to the FISLB, then later, worked for developers as private citizens or even sat on developers' boards. Some politicians, both current and former, developed properties themselves. In 1964, for example, in testimony before a U.S. Senate subcommittee on "Frauds and Misrepresentations Affecting the Elderly," Morton Paulson described a waterlogged subdivision to the west of Daytona called "University Highlands," in which the Firstamerica Development Corporation of Hollywood, Florida, sold raw, unimproved acreage as an "investment opportunity." Unimproved meant uninhabitable—the land was simply too wet, too critical environmentally, or too expensive in terms of infrastructure for building homesites—but Firstamerica and its owners, including former Florida governor Fuller Warren and former state supreme court chief justice T. Frank Hobson, sold it anyway.[14] The price was $1 down, $10 a month.

In the mid-1960s, Bryant's successor Haydon Burns appointed his friend and campaign contributor Leonard Rosen to the board, plus Gulf American vice president Robert H. Finkernagel and "lay" members Howard Hirsch, a former attorney for Gulf American from Miami, and Joseph F. Chapman Jr., Burns's one-time legal aide who had worked with the company as well. Later, when Burns left office in 1967, Rosen rewarded the governor by hiring the firm Haydon Burns & Associates to do consulting work for Modern Air Transport, the Rosen-owned airline company that flew customers to Fort Myers.

To be sure, when Farris Bryant created the board, critics grumbled. It was, after all, a somewhat toothless regulatory agency run by the very companies it was to regulate, but the appointment of Leonard Rosen, not to mention a Gulf American majority in which three of the five board members had Gulf American ties, was excessive even for Florida, where backscratching and patronage were a way of life. When Burns appointed Rosen, in fact, the FILSB had been investigating Gulf American for a host of spurious sales practices at three different developments in central and southern Florida—namely, River Ranch, Remuda Ranch, and Golden Gate Estates.

Like Firstamerica, by the early 1960s the Rosens had decided that, as a business model, providing roads, bridges, canals, and code-compliant homesites was too expensive. According to one 1966 estimate, Cape Coral cost $2,000 an acre to develop, which meant, hypothetically, that if each of its 49,483 acres sold, the company was on the hook for $98,966,000. But, if the company hawked unimproved "investment acreage," with only ditches and a few access roads, unpaved, the cost was just $375 an acre, if that, which is why Gulf American shifted its focus to selling lots in the boondocks.[15] Of course, the prices were lower there, but the volume was higher, and the follow-up costs were nil. "We're promising nothing," explained Leonard Rosen, "absolutely nothing. We're offering what we believe to be an excellent investment at low cost and with the company's good name behind it. . . . River Ranch will be a place where the unsophisticated investor can come for an inexpensive vacation and take pleasure in his property. Then, as the land grows in value, he can do as he wishes with our backing."[16]

The ads for River Ranch were all above board, at least to the FILSB, which judged whether or not they "made a full or fair disclosure" or were "false or misleading" under the law. And who could argue with this blunt 1965 disclaimer, which Gulf American included in its brochure? It read: "[The land at River Ranch] is being offered with no improvement claims as raw, unimproved acreage, and other than State Highways 60 and 630 on its northern boundaries, and the existing farm and ranch roads through the property, there are no roads which provide access to individual tracts, and none are contemplated." As for water issues, it stated that "the existing acreage, while suitable for farming and ranching, does not eliminate ponding and surface water during the rainy season; and no additional corrective drainage is . . . [forthcoming]."[17]

The problem was that the average buyer didn't know that. Buyers had either not read the brochure or even seen it and bought land over the telephone from smooth-talking salesmen who implied or simply told them that

A 1965 advertisement for River Ranch Acres. Although most big developers sold buildable homesites, some, such as the Gulf American Corporation, found community building too expensive, so they instead peddled unimproved acreage in Florida's swamps. Courtesy of the author.

they could live there—that River Ranch was inhabitable, when, in fact, it was not. "They [told me] they were going to build condos" there, said Wasyll Gina, a retired bank auditor from Connecticut, and "that Disney World was going to extend [south to River Ranch and put] . . . a water park in there. We northerners fell for it hook, line and sinker."[18] What Gulf American had sold them, in effect, was an incredibly expensive membership in a hunting and fishing club, where, with the purchase of a parcel of 1¼ or 2¼ acres for

$1,775 to $2,875, buyers received the run of the property, all 43,000 acres, and the right to stay in a lodge (which they paid for, of course) and to use various other facilities, such as a shooting range, a fishing dock, or a dog kennel, provided by the company.[19]

It was a gimmick, of course, something to show and entice buyers during visits and to get them into the "chute" (i.e., the "cattle chute"), which is what salesmen at Golden Gate called the twenty-six tiny cubicles where they delivered their sales pitch. Journalist Trevor Armbrister described the experience in a 1963 article in the *Saturday Evening Post*. A "two-lane highway from Miami led through the foreboding Everglades," he wrote. "To the left, we could see nothing but swamp; to the right, interrupted by occasional Seminole Indian settlements, was more swamp. We spotted the first billboard for Golden Gate Estates as we neared the town of Naples. It featured a young lady lying on a beach with her back to the camera. The caption read: 'You ain't seen nothin' yet.'"[20] As with River Ranch, Golden Gate Estates consisted entirely of investment acreage, over a hundred square miles of it; however, the attraction here was that, on a neighboring tract, several miles to the north and west, Gulf American had platted a planned community called Golden Gate.

It had streets and homesites but was just four square miles; therefore the pitch, discovered Armbrister, was that if he bought land in the uninhabited Estates area, as Golden Gate developed and became a city, his two-and-a-half-acre parcel would be eight suburban lots, which he in turn could subdivide and sell, a seriously sound investment. Never mind that perhaps half of the Estates subdivision, specifically the southern part, had been built in a clear-cut, former cypress forest called the Picayune Strand, which was critical to the Everglades ecosystem but so waterlogged and rugged that one person called it fit only "for movie companies making films about escaped convicts."[21]

So why did people buy unusable, unimprovable, so-called investment acreage in the midst of a Florida swamp? After all, this wasn't Cape Coral, where vacant but still-buildable homesites stretched far into the horizon and where residents sometimes had lots with ocean access and even a yacht club. These were the Golden Gate Estates, a stretch of wilderness in the western Everglades, where panthers ate livestock and alligators lived in sinkholes and where, Susan Orlean wrote in her 1998 bestseller *The Orchid Thief*, "every mile looked like every other mile . . . profuse and green and impenetrable." Orlean also wrote of a man chasing a black bear while armed with an assault rifle, and of a mysterious Bigfoot figure that the locals called "Skunk Man."[22]

The point is, only the truly misinformed bought land here sight unseen. Yet others were greedy too, and covetous, taken in by assurances that this was "the ground floor" and that all land was unimproved land *prior to development.*

"What the land companies did best," wrote one observer, "was appeal to the customer's greed and willingness to be cut in on a sure thing." They were very good at giving "the impression that the real estate potential of 'Blue Slough Highlands,'" for example, "was just a few years behind" Miami Beach.[23] However, when payments mounted and profits failed to materialize, buyers complained. They complained so much that by the mid-1960s, some two dozen states (and even Canadian provinces) had passed or strengthened installment land sales laws, making it a requirement that developers who sold land in those states register, submit property reports, and be approved by a state agency. Then, in August 1966, a U.S. Senate subcommittee held hearings on S. 2672, the proposed "Interstate Land Sales Full Disclosure Act," which required developers to register their properties with the Securities and Exchange Commission and made land fraud a federal crime.

With testimony from nineteen different witnesses, including journalist Morton Paulson and the vice president of the National Better Business Bureau, the subcommittee focused its attention on companies that sold raw investment acreage and that, by all accounts, were the worst-of-the-worst: Rocket City, where developer Milt Mendelsohn sold lots that were 75-feet long and 600-feet wide; Klamath Falls Forest Estates in Oregon, where buyers were shocked to discover that installing water wells required pipes that went 900 feet down; and Golden Palm Acres near Naples, whose "high and dry" homesites were not only not homesites but part of a state flood control district that was rarely, if ever, dry.[24]

In one exchange, subcommittee chairman Senator Harrison Williams of New Jersey asked Stanley Selig, president of the Selig Bros. firm of Indianapolis, to explain a promotional brochure he had issued touting investment acreage in Brazil. On the cover were color photos of architect Oscar Niemeyer's glistening capital of Brasilia, with the caption: "An invitation to meet at the foot of the rainbow to divide the pot of gold." But, when Williams asked Selig how far Brasília was from the acreage, Selig replied, "About four hundred miles." As a reporter for the *Tallahassee Democrat* wrote, "Williams stood up, gnawed on his spectacles, and fixed Selig with a baleful gaze." But "that's the whole front of the brochure!" exclaimed Williams. "I would think it had something to do with the property. Can you give us any justification for this?" Selig replied, "I don't think it needs it, sir. That's my business."[25]

The most telling bit of testimony, perhaps, wasn't testimony at all, but the dramatic no-show of Leonard Rosen, who had been slated to appear. Citing "airline problems" as well as "urgent business matters requiring [his] attention," at the last minute Rosen skipped his appointment—instead sending written answers to questions and letters of support from a pastor, a rabbi, and a priest, no joke—which prompted an annoyed Williams to state: "Gulf American is probably the world's largest installment land sales and development company. Approximately 90,000 people are now paying on property they have bought [under its auspices]. In addition, Mr. Rosen is a member of the Florida Installment Land Sales Board." And, since "several critical comments had been made" at these hearings about his membership on the board, "we thought it important that [he] be asked to testify in person." Now, "we don't subpoena here. This committee is gentle in its legislative processes. [But] I don't think we'll close this hearing. We will give him another opportunity to come in."[26]

Rosen never did come in; his lawyers told him not to. But, by failing to testify before Congress, he opened a Pandora's box that even he and his political ally Haydon Burns, the governor who appointed him to the FILSB in the first place, would struggle to close. With articles critical of Rosen in newspapers across Florida, including one whose headline proclaimed "Rosen Cops Out as Witness at Mail-Order Land Probe," residents wondered why such a shady developer would even be on the board, let alone its chairman, and if he had something to hide. As it stood, Gulf American did 25 percent of all installment land sales in Florida, but through the first half of 1966, it accounted for 51 percent of all FILSB complaints. Its size was staggering. In 1966, for example, its 5,000 employees did some $144 million in sales, about $1.1 billion today, earning more than $22 million in profit. The company owned 300,000 acres in Florida and another 100,000 acres elsewhere in the United States, had its own airline, and threw 12,000 cocktail parties a year.[27] It also owned the "Gulf American Galleries" in Miami, a collection of rare artworks by Picasso, Degas, and Rodin, among others, and in 1965 it tried but failed to purchase the pro football expansion franchise that became the Miami Dolphins.[28]

The complaints were sprawling. Salesmen had promised homesites where none were available; they promised ocean access or had guaranteed profits, or had withheld property reports, or had said they would buy back lots, at a markup, when customers were ready sell. On more than one occasion, Gulf American sold property it didn't own (but held options to) and knowingly sold lots that hadn't been registered with the state. The company also engaged in a fraudulent practice known as "lot switching," which is the reassigning of lots from one location to another without telling the buyer. It did this on at least two different occasions, most notably in Cape Coral, where workers clearing homesites discovered rich deposits of marl, a mixture of calcium carbonate and other minerals used in the making of roads. Rather than pay for the marl, the Rosens simply stole it. "No one knows how the secret was uncovered," wrote one employee, "but it was said that a man who came to see his homesite" discovered that the lot he had purchased had somehow been "swallowed" by a hole. "He was sent to Miami, put up in a hotel and given elaborate treatment, but the secret got out anyway."[29]

The lot switching complaint was a big one. But, like other complaints brought before the FILSB, those involving land giants, and Gulf American in particular, had a habit of disappearing following "minor corrective action" or simply were ignored. As one person put it: "It was seldom anything you could put your finger on, but in the time" Gulf American controlled the board, hearings into its activities "just never seemed to get anywhere. Somebody always was sick or tired, or had to catch an airplane."[30] On occasion, board member Elliott Mackle complained to the press, albeit mutedly, insisting that the FILSB, in its present form, was "inadequate" at curbing abuse. He didn't name the Rosens but railed against raw acreage developments, claiming swamp sales were "killing Florida," and that if state or federal officials didn't do something, they'd experience "another 1926."[31] That was the year of the last property meltdown, when Florida went broke.

Indeed, the Mackles' Deltona Corporation differed from the Rosens' Gulf American Corporation in that it only sold homesites—"We won't touch a piece of property if we can't put good homes there," said Elliott. But, in their on-site sales pitches, both the Mackles and Rosens played a similar game. "As we see it, there are two kinds of developers" in Florida, said the head of the Office of Interstate Land Sales in the Department of Housing and Urban Development. "The guys who use high pressure techniques, but keep their promises. . . . Then there are the guys who use high pressure salesmanship to defraud and swindle."[32] Although a simplistic description, to be sure, the Mackles, by most estimates, belonged to the first group while the Rosens belonged to the second. An acquaintance of both sets of brothers put it a different way: "Sure, they both sold land, and both used questionable tactics in selling and transforming the land, but the Mackles were what I'd call conventional businessmen. They were gentlemen. But the Rosens? The Rosens were rogues."[33]

Based in Baltimore, the younger of the two Rosens was Jack, a highly intelligent, somewhat introverted man in his forties who suffered from a heart condition. Bookish yet sociable, Jack ran everything at Gulf American that wasn't in Florida, including sales dinners, corporate relations, and marketing, while Leonard handled things on the ground. They fought constantly, and while Jack was affable, the fifty-year-old Leonard was self-centered and crass. Gulf American executive Leonard Rich worked for both brothers, first with Jack in Baltimore, then with Leonard in Miami. "Jack had scruples," he explained. But "Leonard really had no scruples whatsoever. He was an arrogant guy. He loved to flout people" and "didn't give a damn what his [salesmen] did."[34] The elder Rosen cursed like a sailor and often used the bathroom while dictating directions to a secretary who sat outside. He had no problem changing and stripping naked in front of his executives and once held a business meeting with no shirt on while eating buttered popcorn and wiping his hands on his chest.

Dynamic, impulsive, hard-driving, and volatile, Leonard was "a splendid example of male chauvinism, flaming egotism, and impressive sensuality," wrote Eileen Bernard, and "like Freud, he had 'examined himself thoroughly and come to the conclusion that he didn't need to change much.'" The neurotic Jack wanted to be him—or at least compete with him—which led to what Bernard called the "world championship of sibling rivalry."[35] As a result, the company had two staffs, two hierarchies, and two headquarters, with Jack favoring the building of additional planned communities like Cape Coral and Leonard the selling of unimproved acreage like Golden Gate Estates. Leonard won the argument, but it was Gulf American's Golden Gate adventure that got the company into trouble. It probably wouldn't have, but in May 1966, Haydon Burns had been upset in the state's Democratic gubernatorial primary, losing to Robert King High.

Although a Democrat in a state full of Democrats, High was incredibly vulnerable. For one, he was an anti-segregationist liberal, a so-called northern liberal, at a time when millions of white Southerners, fearful of the Civil Rights Act and the end of Jim Crow laws, began leaving the Democratic Party in droves. In 1964, for example, Republican presidential candidate Barry Goldwater, who consciously opposed civil rights legislation in order to attract Southern voters, won Georgia, Alabama, Mississippi, South Carolina, and Louisiana, and almost won Florida. That year, Goldwater received some 49 percent of the vote in Florida, a remarkable showing for a Republican in a state where Democrats outnumbered Republicans 1,802,230 to 396,885.[36] A second flaw in High's candidacy was that the forty-two-year-old lawyer hailed from Miami, a city some Floridians, but especially those in northern and central Florida, considered Sodom and Gomorrah. "To them," it was "something alien," wrote historian Edmund F. Kallina, and though "few would say it aloud, most saw Miami as controlled by Jewish interests." It also had a large African-American population that Burns called the "Negro bloc vote," and was, as Claude Kirk put it, "a citadel of crime."[37]

As Republicans, the Mackles had supported Kirk, but no one anywhere in Florida thought he would win. That is until High beat Burns in the primary, and Kirk, a law-and-order, pro-business Republican from Jacksonville, came to the fore. "We were having a primary night party for Kirk at my house," said campaign adviser Nathaniel P. Reed, "and my wife said, 'Hey! Let's turn on the news to see who won with the Democrats.' She was joking, of course, because Burns was such a shoo-in everyone laughed. But, the story was it had rained all over northern Florida that day, while in South Florida it was sunny and bright. The turnout favored High and High won. We were speechless. I mean everyone was quiet. That's when I turned to Kirk and said, 'Congratulations, Claude. You're the next governor of Florida.'"[38] While shocking to some, for the Mackles, the emergence of Claude Kirk was the opportunity they'd been waiting for. By backing the Republican financially, he would do for them what three different Democratic governors had refused to do—that is, empower the board.

Although sales had been strong in Florida for each of the land giants and especially Gulf American, through 1966 complaints had risen, substantially, and both the Police Benevolent Association and the United Firemen's Association had warned members not to buy Florida land. Perhaps most damaging, however, was an August 1966 column, syndicated nationally, by the respected and plain-spoken economist Sylvia Porter, author of such popular how-to books as *Managing Your Money* and *Money and You*, in which Porter wrote: "Don't buy land you have never seen" and of "which you know nothing [about]" because "mail order land peddlers" are "reaping huge profits on useless desert and swampland from coast to coast," and though states like Florida have "enacted new laws to control such sales . . . much of the $1 billion in Florida real estate now up for [purchase] is uninhabitable." The column's headline? "Beware Mail-Order Land Gyps."[39]

Clearly, officials in Florida needed to do something before the entire industry collapsed. But when Leonard Rosen resigned from the FILSB in September 1966 following his no-show with the Senate, the pliant Burns replaced him with company associate Joseph F. Chapman Jr. That continued the three-to-two Gulf American majority, and though FILSB investigators had worked up a damning report on the company, one that detailed, among other things, "grossly immoral and unethical sales activities," including 1,300 new instances of lot switching, it wasn't publicly heard.[40] That is, until Claude Kirk became governor and leaked the report to the press.

ON SATURDAY, 10 DECEMBER 1966, an East Naples, Florida, resident named Clayton Alvin Hamm had a late-night argument with a drunk and belligerent cousin named Randy Hamm that ended when Clayton killed Randy with a .357 magnum.[41] In what was an open and shut case, police arrested Clayton for murder for pumping six holes in his cousin through an open bedroom window. But, according to Florida law, prosecutors could only try Clayton for murder following an indictment from a grand jury. That jury, made up of eighteen Collier County residents, including a carpenter, a bookkeeper, and a cashier, was impaneled three days later, on 13 December 1966, by circuit judge Harold S. Smith, who told jurors "the searching eye and inquiring mind of a grand jury is the greatest deterrent to evil and corruption we have."[42] Although brought together for the Hamm case, which was its only case, legally the jury could hear evidence of any crime, not just murder, and could investigate, and suggest remedies for, "how public officials [were] conducting their offices and discharging their public trusts."[43]

No one, apart from lawyers and court junkies, really knew this, but a needling, corruption-fighting, sixty-five-year-old tax critic and real estate broker named Vince Conboy did. Conboy had been at war with Naples for years, claiming that the sons and heirs of Barron Gift Collier, a wealthy advertising magnate who acquired, by the 1920s, over a million acres of Southwest Florida land, had been given what he called "a developer discount" by tax accessors, who had valued lots in Collier's Port Royal development at less than market value.[44] In 1965, Conboy took the case to the Florida Supreme Court and won, then in 1966, at the Hamm grand jury, he and three other men, including two former Gulf American associates who had worked at Golden Gate, asked foreman George Huntoon if they could testify. They had evidence of wrongdoing, they claimed, and though Huntoon was vice president of the Collier Development Corporation and though the judge's former partner was an attorney for Gulf American who had done business with the Colliers, there was nothing Huntoon could do. The men were allowed to testify.

Although given in private, the evidence they presented was well known. Gulf American had used deceptive sales tactics, had failed to show buyers property reports, had engaged in lot switching, and had knowingly told customers that soon they would be flipping their land for a profit. That last point was patently untrue, and Conboy, whose Gulf Coast Realty office had received hundreds of letters from ripped-off customers hoping that he, a broker, would get their money back, could prove it. "I feel an obligation to answer these letters," he told one buyer. "Hundreds of people all over the United States are buying this acreage and most of them [are as] uneasy about their investment as you are." But "the only people in Collier County who feel that this land has value . . . are the men operating Golden Gate Estates." And "you and I both know that they wouldn't give you a dime for your equity," so "quit trying to get rich" and please "seek advice on your real estate investments in the future."[45]

Once, when Conboy told a Virginia woman that the property she owned had no resale value whatsoever, Gulf American counsel Harry Warren threatened him with legal action, declaring that if Conboy made any more remarks "in a disparaging manner against Gulf," then Warren would do whatever was legally necessary "to protect the company's interest." He warned Conboy, "Please govern yourself accordingly." Conboy, however, was unmoved by the threat and responded: "Thanks for your letter of May 16th. I am answering promptly so that you may exercise your further actions without delay. Apparently we don't agree as to the 'Market Value' of the land you are selling in Collier County. A good place to determine the actual value of the land would be a 'Court Room' with men under oath. I'm sure this information would be valuable to the hundreds of your buyers who have swamped local brokers . . . [hoping] to liquidate their equities."[46]

Wisely, Warren never contacted Conboy again, and though the Hamm grand jury didn't indict anyone except Hamm, in January 1967, it issued a report (or a "presentment") in which it suggested that Governor Claude Kirk reorganize the FILSB to better police land frauds. Calling it an "issue of local and statewide concern," the jury recommended a five-member board, in which the first four members were (1) a banker, (2) a mortgage broker, (3) a real estate agent, and (4) a title attorney, with no "past or present affiliation with land developers." Then, to assist the board with understanding the industry's "particular problems," the fifth member would be a developer.[47] Thus, to recap: The jury's plan was for four unaffiliated members and one developer—not the indictments or jail terms that Conboy had been looking for. Still, it was, as the Associated Press put it, "a slap" at the FILSB, one that prompted Kirk to push for legislative action, in the spring of 1967, to abolish and reconstitute the board.[48]

No doubt, had he had won, Governor Haydon Burns would have ignored the issue, but Kirk was no ordinary governor and Florida was no ordinary place. For one, the state was expanding, exponentially, adding some 5 million new residents between 1950 and 1970. (That's 4,700 move-ins a week or 245,000 a year.) Most would settle in central and southern Florida, where, by sheer force of numbers, they would upset what many argued was the most unfair political system in the country, in which each of Florida's sixty-seven counties had one senator and a maximum of three representatives *regardless of population*. As a result, the imbalances inherent in the system were amazing. According to one scholar, in 1960, for example, Dade County had over 900,000 people but the same number of house representatives, three, as rural Lafayette, Liberty, and Glades Counties, which had one each. With just 8,977 residents combined, they were a mind-boggling 104 times smaller, population-wise, than Dade.[49]

However, unequal representation was one thing; states in the South had a long history of it. But in Florida, a group of senators from north and west of the Suwannee, mostly from flea-speck towns like Quincy and Bonifay in wooded regions of the Panhandle, banded together in a political bloc, whose conservative, anti-integrationist, and anti-urban ruling philosophy governed the entire state. Dubbed the "Pork Chop Gang" by an editor at the *Tampa Tribune*, in the 1950s, the bloc had at least fourteen members, and "it was clear," wrote one observer, "that they were in charge of the Senate. From their ranks came the President. From him came to them plums of committee assignments," and fourteen votes were "more than enough to block the two-thirds required to bring up a bill for consideration." And, "on roll-call after roll-call, the same names showed up against reapportionment, against constitutional revision," and "against urban renewal." The gang "was, and is, primarily a defensive clique, getting in its hardest licks against new legislation."[50]

Aligned with the Pork Chop Gang were politicians from small, rural counties in central and southern Florida who joined Pork Chop legislators in creating what, in effect, was a small county caucus, a rural caucus, which, because of the state's inequitable electoral system, needed just 12.3 percent of the vote to control the senate and just 14.7 percent to control the house.[51] Their opponents, the so-called Lamb Choppers, were politicians from Florida's urban counties, specifically Dade (Miami), Orange (Orlando), Hillsborough (Tampa), Pinellas (St. Petersburg), Palm Beach (West Palm Beach), and Broward (Fort Lauderdale), who in 1960 represented over half of the state's citizens who, in turn, paid two-thirds of its taxes. Nevertheless, the Pork Chop Gang had total control, meeting each year with lobbyists and key business leaders at a fish camp in remote Taylor County, where they set the state's legislative agenda and initiated a Red Scare–era witch hunt of communists in the National Association for the Advancement of Colored People (NAACP) and gays in the educational system.

They also stole, or "reapportioned," a huge slice of the state's gas tax money, which the legislature divvied up on a per-county basis according to a formula factoring size, population, and length of roads built before 1931, when Lamb Chop counties like Broward had just 20,000 residents. By 1960, though, Broward had over 300,000 residents, while its massively expanded but still underfunded road network was among the most trafficked in the nation. "The happy drivers of fifty-six [Florida] counties get more money for roads," griped the *Fort Lauderdale News*, "than they pay in gasoline taxes." In fact, one north Florida county received $40 in road revenues for each dollar that it paid.[52] Now add tax money from dog and horse tracks and jai-alai venues operating mainly in the south but which the state split sixty-seven ways, and Pork Chop counties like Liberty in the Panhandle could pay all of their governing expenses while levying no local taxes whatsoever.[53] It was, for Florida's Lamb Chop counties, a raw deal.

The turning point came in 1964, when the United States Supreme Court voted eight to one in the landmark Reynolds v. Simms case that state legislative districts must correspond, as closely as possible, with the principle of "one person, one vote." As Chief Justice Earl Warren opined, rather famously: "Legislators represent people, not trees or acres. Legislators are elected by voters, not farms or cities or economic interests." And, while Pork Chop politicians tried to either waylay reapportionment or twist it to their own ends, the Florida legislature eventually passed measures, amenable to the courts, in February 1967, just one month after Claude Kirk took office.

Tireless, flamboyant, hard-driving, and incorrigible, the Republican Kirk had won the governorship by over 150,000 votes. But he came to power not because of his policy positions, which were conservative, but because of staunch opposition among Democrats to the supposedly super-liberal Robert King High. In fact, for much of the campaign, Kirk spoke mostly in what one newspaper called "broad generalities that a knowing voter can hardly pin down."[54] There'd be no new taxes, or he'd improve schools or be tough on crime, but when journalists asked for specifics, he referred them to a series of "white papers" that hadn't been written yet. They came out near the end of the campaign in September 1966 and, though criticized for being both "voluminous and vague," covered everything from law enforcement, education, and the economy to conservation and the environment. None of it mattered, for Kirk zeroed in on two key issues: High's supposed ultraliberalism and crime. "Kirk used the word [liberalism] like a sledgehammer to crush the Democratic candidate," whose once fine city, said Kirk, had "earned the unenviable reputation as the number two . . . sin city in the nation."[55] Voters agreed, and hundreds of thousands of Floridians, who heretofore had swung Democrat but now were "Demokirks," carried the politician into office.

Citing the Hamm grand jury, Kirk moved quickly to reconstitute the FILSB and, in January 1967, asked each of its five members, including Elliott Mackle, to resign. Mackle said yes, as did Gerald Gould, but the board's three Gulf American apparatchiks—Howard Hirsch, Robert H. Finkernagel, and Joseph F. Chapman Jr.—said no. It was a ploy, of course, by Kirk, who, as governor, could appoint but not remove the board, so he left the FILSB as is, while pushing the state legislature to act. In June, and with little debate, the legislature gave the governor what he'd been looking for: a reconstituted and renamed "Florida Land Sales Board" (FLSB) made up of five members unaffiliated with the industry and just two developers. In addition, it strengthened the law so that salesmen themselves could be suspended, as well as fined, and made officers whose companies engaged in "persistent, willful violations" liable for up two years in prison and a $100,000 fine. "This is a good, strong law with lots of teeth," said Elliott Mackle. "The old one is mostly an advertising law, while this governs the land and the people selling it."[56]

The problem for Kirk, though, was that the "Uniform Land Sales Practices Act" wouldn't take effect until 1 August 1967, which meant Gulf American and its three FILSB board members had over a month to meet, hear charges, and render what everyone knew was a slap on the wrist. But Kirk had other plans. He had read the FILSB's shocking but confidential investigative report on Gulf American's sales practices, so rather than let the board bury it, which is what the Rosens wanted, he allegedly had James Wolf, his press secretary, leak the report to the *Wall Street Journal*, which on 28 June 1967 published a front-page article entitled "Furor in Florida: Land Sales Practices of Gulf American Corp. Touch Off Controversy." In it, journalist Ken Slocum summarized the report, which, as he put it, accused the company of "flagrant misrepresentation in sales of land, of selling lots bearing specific numbers and then switching the lot numbers to other land," and of hiding "key information" from the board.[57]

Although Slocum refused to say who had given him the report, he highlighted the role of Claude Kirk "in fighting to change the makeup of the board" in hopes of freeing it from the Rosens. They were "scoundrels," said Kirk, and while "most of the industry" was "good and the companies honest," he explained, he just couldn't allow the FILSB to engage in "a whitewash." That had been Gulf American's plan, implied Slocum, because "the current board, which [had] shown no desire to consider the [report] previously, now all of a sudden wants to handle the proceedings itself." Caught in a cover-up, Leonard Rosen would tell Slocum that the charges were overblown. For one, Gulf American was Florida's largest land company, he argued. Therefore, such a big operation naturally had more complaints. And, no matter what Kirk said, the company had no undue influence over the board. As for lot switching, Rosen wouldn't comment on Cape Coral, but insisted the company had relocated lots at Golden Gate Estates because the land it had sold there, the company realized once work had commenced, had been in unbuildable sections of swamp.

Why the company had taken three and a half years to tell buyers that, or why it had waited until after the Hamm grand jury to inform them of it, Rosen didn't say (although the reason, most likely, was oil, which Gulf American had been looking for on land it had already sold in the western Everglades).[58] However, what Rosen did say was that the FILSB had targeted Gulf American unfairly, at the behest, presumably, of its competitors, who had supported Kirk. "It's no wonder," he said, spreading his hands above his head. "The board spends all of its time investigating Gulf American, and none at all on Deltona, General Development, and the others. They have a file up there this big on Gulf American, but just little bitty ones on the other companies. We may soft-soap customers to death, but we didn't get this big by beating customers over the head." Nevertheless, the company had lied to them, giving them lots they didn't buy and that had been "switched," and overexaggerated the "investment potential" of land that Vince Conboy and other local realtors had pegged at zero.

The news wasn't new. Gulf American had been censured by the board before, albeit gently, while tales of its land shenanigans had appeared in the press since at least 1963, with Robert Caro's *Newsday* series on installment land sales and reports by Haines Colbert of the *Miami Herald* and Trevor Armbrister of the *Saturday Evening Post*. But Slocum's "Furor in Florida" article was different, because the soon-to-be-installed FLSB was different, less malleable, and when it took office in August 1967, it had the power—and the political backing—to put the company out of business. Investors took note, rushing to dump shares of Gulf American stock as soon as the news broke. It wasn't a panic per se, but the American Stock Exchange (AMEX) fielded so many "sell" orders that it suspended trading in company shares for two days. When it reopened on Friday, 20 June 1967, its stock plummeted from $11.75 a share to $9.00 a share, a loss of nearly 25 percent.[59]

The company received even more bad news in July, when Kirk announced he was sharing Gulf American's FILSB file—and its copious consumer complaints—with Governor Jack Williams of Arizona, where Rosen was planning yet another Milt Mendelsohn project called "Rio Rico" on a 55,000-acre ranch near Nogales.[60] Then, Kirk told FILSB chairman Gerald Gould he couldn't hold meetings anymore; so in August 1967 the FILSB finished its term and disbanded, and the new seven-member FLSB took its place.[61]

As appointed by Kirk, the FLSB's two industry members were Elliott Mackle and Joseph Taravella, president of the land giant Coral Ridge Properties Corp., which, in the early 1960s had built the planned community of Coral Springs in Broward County and a string of condominiums on the "Galt Ocean Mile" in Fort Lauderdale. The five lay members included two attorneys, one of whom worked for the Disney corporation (which had announced plans for its Orlando-area theme park in November 1965), and an accountant, an insurance executive, and a real estate appraiser. On 8 August 1967, they held an organizational meeting in Tallahassee that was attended, briefly, by Kirk, who rolled in after the meeting had started, gave a quick speech in which he told the board that "the era of 'buyer beware' [was] over," then left. He was in and out in five minutes, but because the FLSB handled its business in private, Kirk gave copies of his comments to the press. There was nothing unusual there, except Kirk's staff had handed out the wrong copy, which, on the back, had a note reading: "Prepared by Mr. William Safire. Suggested for use before the Land Sales Board 8-8-67."

Safire was *the* William Safire, the Nixon adviser, professional publicist, and future *New York Times* writer, who had been hired by Kirk, not without controversy, to "use the governor's image nationally in a program to attract industry to the state"—in other words, to put Kirk on Nixon's radar as a viable running mate for the 1968 campaign.[62] Kirk had given Safire a whopping $90,000-a-year contract through the Florida Development Commission, an exorbitant amount worth nearly $700,000 today, and though Safire had assured politicians he'd be promoting Florida, he was actually promoting Kirk. "The main purpose is to boost Florida," said one insider. But "if a byproduct is to boost Governor Kirk, then that is alright too."[63] Working behind the scenes, Kirk had landed, for Florida, the 1968 Republican National Convention and planned a "barnstorming for industry tour" in November 1967 that included Safire-planned stops at GOP fundraisers in Wisconsin, Illinois, and Ohio, plus a coveted appearance on *The Tonight Show*.[64]

Kirk took Safire's advice on everything, it seems, from how to deal with Alabama Governor George Wallace's third-party candidacy to the Jacksonville visit of H. Rap Brown to remarks given, in a seemingly inconsequential speech, before the Land Sales Board. But the issue was important to Kirk. The Mackles were friends of his, not to mention donors, plus earlier that year Nixon had opened a golf tournament at the Mackles' Marco Island Country Club and had purchased the first of several properties on Key Biscayne.[65] In fact, Nixon's future neighbor there was the banker Charles G. "Bebe" Rebozo, a longtime friend and confidant who lived in a Mackle-built home. Nevertheless, Kirk's insistence that the FLSB investigate the Rosens was more than a political favor because, as virtually anyone who ever opposed him learned, the governor was pathologically confrontational. His "politics were visceral," wrote one historian. "He reacted with a vengeance, sometimes petty, against individuals, institutions, and organizations that 'annoyed' him."[66]

Suffice it to say, Leonard Rosen annoyed him, for the wily developer had not only opposed Kirk in the election, aggressively, but had refused a request by the newly installed FLSB to make its investigation public. Although Florida had recently enacted its so-called Sunshine Law requiring government meetings at both the state and county level to be open to the public, the ruling didn't apply to investigations, so at least in the short run, his company was off the hook. Yet somehow, in September 1967, Rosen managed to antagonize an agency far greater than the board, in this case the National Aeronautics and Space Administration (NASA). The space center was stunned to discover that Gulf American had built a reception center south of the NASA Causeway in Titusville and had been giving away free tours. While not technically illegal—Gulf American had paid for the tickets like everyone else—it had collected NASA-made brochures from local hotels and tourist establishments and printed "free space center tours" with the company's number on the back. Then, it flew customers to River Ranch, a 114-mile trip with a film and a sales pitch, before giving them their tickets.

For a time, Gulf American moved upwards of seventy to a hundred visitors a day through what it called its "Two-for-the-Price-of-None" tours, but it curtailed the practice when the space center's public affairs office complained.[67] Rosen had enough to worry about. In October 1967, the FLSB was set to examine the Gulf American case and, with Kirk claiming the board would "protect Florida from the deprivation of high pressure salesmanship and unscrupulous businessmen," things didn't look good.[68] So Rosen, in a move clearly meant to goad Kirk, hired a Tallahassee attorney named Mallory Horne, a sitting state senator and Democrat, who on several key issues, including money for state capital renovations and education, had opposed Kirk. Naturally, the governor was apoplectic and pilloried Horne for having a "carefree lack of ethics," but Horne claimed he'd reviewed the company's books and found no instances of "fraud against the . . . purchaser." Therefore, he was assisting Gulf American's chief legal counsel in finding "an area to stipulate" an end to the investigation before the board.[69]

The move was pure Rosen: hiring an insider and Kirk opponent who just happened to be one of the most powerful men in the senate. (In fact, Horne had recently outmaneuvered urban Democrats in scuttling a very real plan to move the state's capital to Orlando.) It was brazen, brilliant, and shortsighted. Leonard "was arrogant as hell," said former Gulf American executive Leonard Rich. "With any regulatory agency, state or federal, when they have a job to do they normally go after the biggest guy. If they could control Gulf American and knock us on our ear, they'd be hitting a substantial part of the whole industry." But "Leonard thumbed his nose at them." Gulf American was "the prime target, and of course he managed to antagonize them." Nevertheless, Rich insisted that the FLSB "would have softened the blow" if Leonard hadn't carried such "a big chip on his shoulder" and hadn't been daring them to "knock it off."[70]

But the feisty developer couldn't help himself. He excoriated the board for being biased, not to mention incompetent, and once, in a fit of rage, told the *News-Press*: "They're bums! All of them are bums! They are nothing! You hear? They are nothing, sucking on the blood of those who produce! I am a builder. I am not a taker." Rosen would also claim that Kirk and the Mackles were anti-Semitic, without proof, and that Elliott Mackle and members of the Land Sales Board were conspiring against him. Mackle's "envy," said Rosen, "knows no bounds."[71] But, with the board closing in and rumors now of criminal charges for not just salesmen, but for executives, Rosen went for the jugular, attempting to blackmail Kirk with news that close friend Shelby Dale, an attorney from Fort Lauderdale and Kirk's recently appointed head of the Florida Development Commission, had received bribes and other inducements from various businessmen and had also had close dealings with Gulf American.[72]

As company lobbyist Barry Horenbein insists, Rosen initiated the plan in October 1967, when he learned that the FLSB intended to hit Gulf American with a "show cause" order asking it to answer, in writing, why its license shouldn't be revoked. Horenbein, who lived in Tallahassee and represented companies like 3M, received a late-night phone call from the developer, who, agitated and angry, told the lobbyist what he had. "Now, if the governor is getting ready to play hardball with me," he said, then "I'm prepared to play hardball [with him]. They're trying to put my company out of business,

Barry, for the good of Mackle. . . . [So], I'd like you to go in . . . and tell them . . . if they still want to hold their press conference," then "I'm going to hold mine. And I'm going to tell them about Shelby Dale, and I'm going to tell them about this person, and I'm going to tell them about that person, and I'm going to tell them about how much they asked for in the governor's race and why they asked for cash. . . . I'm ready. I've had a year of this shit and I'm ready."[73]

Following orders, Horenbein called the governor's office and set up a meeting but insisted he hadn't pressured Kirk or told him what to do. He simply informed the governor what Rosen had on Dale, verbatim, and that if the FLSB held a press conference and issued a show-cause order, Rosen would hold one too. However, Kirk did an end-around, promptly suspending Dale and announcing at an impromptu press conference on Friday, 13 October, that Rosen had tried to blackmail him. "Last night a lobbyist retained by Gulf American Corp. approached several members of my staff with a deal," he explained, "that if the governor would . . . take the heat off Gulf American, then Gulf American would not reveal some embarrassing information" it had gathered "involving an appointee of the governor." He then intoned: "You will make no deals with me." Two hours later, the FLSB issued its show-cause order, and though Rosen declared that both he and Gulf American would be exonerated, AMEX suspended its stock temporarily, amidst sell-offs, for the second time in a year.[74]

Cornered and facing jail time, Rosen, his brother, and six company executives had until 10 November 1967 to respond, but wisely struck a deal with the board in which Gulf American pleaded guilty to five separate charges, from lot switching and the re-platting and sale of land without FLSB knowledge or approval, to changing the plan of sale of an approved parcel and engaging in deceptive sales practices. It also agreed to suspend sales for thirty days; to pay for, and give access to, a group of five board-appointed monitors who would oversee the company for 150 days, during which time each of the 1,300 buyers in Collier County whose lots had been switched could apply for a refund; and pay a $5,000 fine.[75] "The days of shenanigans and hanky-panky in the sales of Florida land are gone," proclaimed FLSB executive secretary Carl A. Bertoch, who either was misinformed, overly optimistic, or both. "Florida land can now be purchased with confidence."[76]

Realistically, though, the FLSB had approved the deal because, like Goldman Sachs or AIG in the 2000s, Gulf American was too big to fail. By 1966, for example, it was the fourth-largest company in the state, having had over a 9½-year period an estimated $1.4 billion impact on Florida's economy. Over t same time, it had spent $68.9 million in payroll and commissions and $50 million in development expenditures, and it had purchased, just in Lee, Collier, and Polk counties, $67.5 million in land. What is more, Gulf American owed at least $90 million in improvements, had served (or was serving) 180,000 purchasers and 15,000 stockholders, and had recently acquired an airline company called Modern Air Transport that brought to Florida some 10,000 people a month. The bottom line: Gulf American was huge. It was so huge, in fact, that its one-month FLSB-ordered suspension from 10 December 1967 to 10 January 1968 resulted in layoffs of over 1,000 people and the loss of $5 million to $10 million in sales.[77]

Even so, Gulf American was lucky, and though its five-month monitorship was, as the FLSB put it, "a chance to correct itself without having its license revoked," Leonard Rosen failed to view it as a soft punishment. He stewed through December and in January 1968 announced suddenly that he was suing the FLSB for overstepping its bounds, on the legal technicality that violations committed before 1 August 1967, under the old land sales law, could not be prosecuted under the new sales law and thus not by the board. He then filed a $16 million lawsuit against Kirk; Bertoch; Dow Jones, Inc.; Wolf; and Kirk assistant Tom Ferguson for "conspiring to violate the secrecy provisions" of Florida law for leaking the contents of Gulf American's investigative report to the *Wall Street Journal*. The Kirk suit stretched into the summer, but in February 1968, a Florida court of appeals sided with Rosen, ruling that no matter what crimes he and his company had admitted to, if they'd committed the crimes before August 1, then the FLSB couldn't punish them.[78]

Humiliated, the board retaliated, turning over its Gulf American files to grand juries in Lee, Collier, and Monroe counties (the latter of which was uninhabited except for the Keys but which had jurisdiction over company holdings in the Everglades) and denying the registration of eight new parcels, in total 15,000 acres in central and southern Florida that the company had been planning to sell. Rosen sued again, this time claiming Bertoch, the board, and most notably Elliott Mackle were biased. He pressured Bertoch, a mid-level state employee and the only person named in Rosen's *Wall Street Journal* suit who couldn't afford counsel, by serving *only him* with papers. Rosen had muddied the waters significantly and had proved himself yet again to be a prodigious fighter, even a dirty fighter. But two stock suspensions and a guilty plea, though rescinded, had led to a mountain of bad press. The company had sold no land through December, had its licenses revoked in Pennsylvania and New Jersey, saw Connie Mack Jr. quit, and witnessed an uptick in customer lawsuits, including one in which an aggrieved Rhode Island man asked a federal court in Miami to prohibit the company from advertising through the mail.[79]

In fact, for the six-month period ending on 29 February 1968, sales totaled $44.7 million, down from $63.98 million that same period the previous year. "The adverse results" they'd suffered, said one executive, "were brought about by the year-long controversy and litigation" with the land sales board. As a result, there was a sharp reduction in cumulative sales."[80] With profits down and declining and his parcels in limbo, Rosen resigned from his post as board chairman in early July 1968 and offered the FLSB a deal: He'd drop all pending litigation against the board in exchange for its approval of the parcels and a "clean slate." It seemed like a joke, said a board member from Jacksonville, "like saying you'll stop calling me names if I give you candy."[81] And besides, Leonard and Jack Rosen still controlled the company, owning approximately 60 percent, or $50 million worth, of its 9.5 million shares. The board said no.

The only way out, realized Rosen, was to sell the company, which he did on 17 July 1968 in a $200 million "stock swap" with a Pennsylvania-based financial conglomerate called the General Acceptance Corporation. Under its terms, the Rosens resigned from their positions and were given one-year consulting contracts and limited voting power but received at least $125 million in stock in a Gulf American successor company called GAC Properties.[82] The end in sight, the FLSB and Gulf American buried the hatchet a week later. The company withdrew its lawsuits, including the one with Bertoch and Kirk, while the board ended its investigation and registered the company's land. This was "not a draw," insisted board member Charles W. Johnson. "They cleaned up over there terrifically. They agreed to every refund requested and their financial accounting and projections were greatly improved." They did "a complete about-face," which "is what we wanted them to do to begin with."[83]

In November, Claude Kirk's Florida Development Commission, the agency formerly run by the now-fired Shelby Dale, sponsored a junket, a "plush" fifty-person tour by editors of trade and financial magazines to sites throughout Florida, from Cape Kennedy, Fort Lauderdale, and Orlando to Miami and Cape Coral. The Commission's plane was a brand-new Convair 990 Coronado, which had been donated for the trip by Modern Air Transport, a subsidiary of GAC Properties.[84] Obviously, the Rosens were gone, but the company and its salesmen were back in business. As for the Mackles, the outcome of the FLSB–Gulf American regulatory fight was more or less what they'd wanted: to clean up a swamp peddler, whose relentless, sometimes questionable sales tactics had sullied an entire industry.

And now, with the Rosens gone, the Mackles were the most powerful developers in the state. In many ways, they were the state, for as one banker joked: "I can tell when we leave Georgia and start flying over Florida: I hear Mackle hammers on Mackle roofs. Sometimes I think I can hear Mackle fingers counting Mackle money."[85] However, by the mid-1970s, a world oil crisis, a stock market crash, and a recession would cripple their company, which lost $6.5 million in 1974 and $5 million in 1975 before the unthinkable happened: Citing environmental issues, the Army Corps of Engineers killed its project at Marco. And, when it did, Deltona lost some $50 million in a day.

CHAPTER FOUR

# The Last Paradise

> With no mountains or gorges, Florida has no natural impediments to man-made changes. Florida is like a big bag of Silly Putty, and it is being reshaped irreversibly in the hands of big subdivision developers.
>
> —JOEL KUPERBERG, Internal Improvement Fund of Florida

In the southernmost reaches of Southwest Florida, where scrub brush and pineland turn gradually into swamp, sits Collier County, the richest county per capita in the state. As of 2017, there were six billionaires in the Naples–Marco Island Metropolitan Statistical Area, where an estimated 8.6 percent of residents, or 27,500 people, possessed a $1 million or more.[1] Collier was founded by a millionaire, the advertising magnate Barron Gift Collier, who, between 1921 and 1923, acquired a mind-boggling 1.3 million acres of Florida land. Some 900,000 of those acres were in southern Lee County, which at the time was the largest county in terms of land in the eastern United States. But Collier wanted more. He wanted not just land but his own administrative district, so in 1923 he promised to spend $10 million to develop the region if politicians in Tallahassee created Collier County. It was an audacious plan, opposed by Lee County residents in particular, but no one but Collier had the money or the wherewithal to build a port in the Everglades, for example, or an especially tricky, thirty-one-mile stretch of the Tamiami Trail.

"[Let it be] said that Barron Collier is 'sold' on South Florida," wrote an op-ed in the *Lakeland Star-Telegram*. "He lives in it and believes in it and boosts for it," and "is said to have worked up some wonderful plans for the development of the broad acres now in his possession. . . . [If] the legislature will only make . . . [him] King Collier of Collier County, he proposes to show us the real meaning of that oft-used and oft-abused word, development."[2] Collier, who had made his fortune in print advertisements used in trollies and subways throughout the United States, had his work cut out for him, for as of 1920, there were just 1,224 people in total living in what became Collier County. Approximately one in five was a Miccosukee Indian. At the time, they were members of the Seminole Tribe living in thatched huts in the Everglades. There was a 100-person cattle community at Immokalee, which was

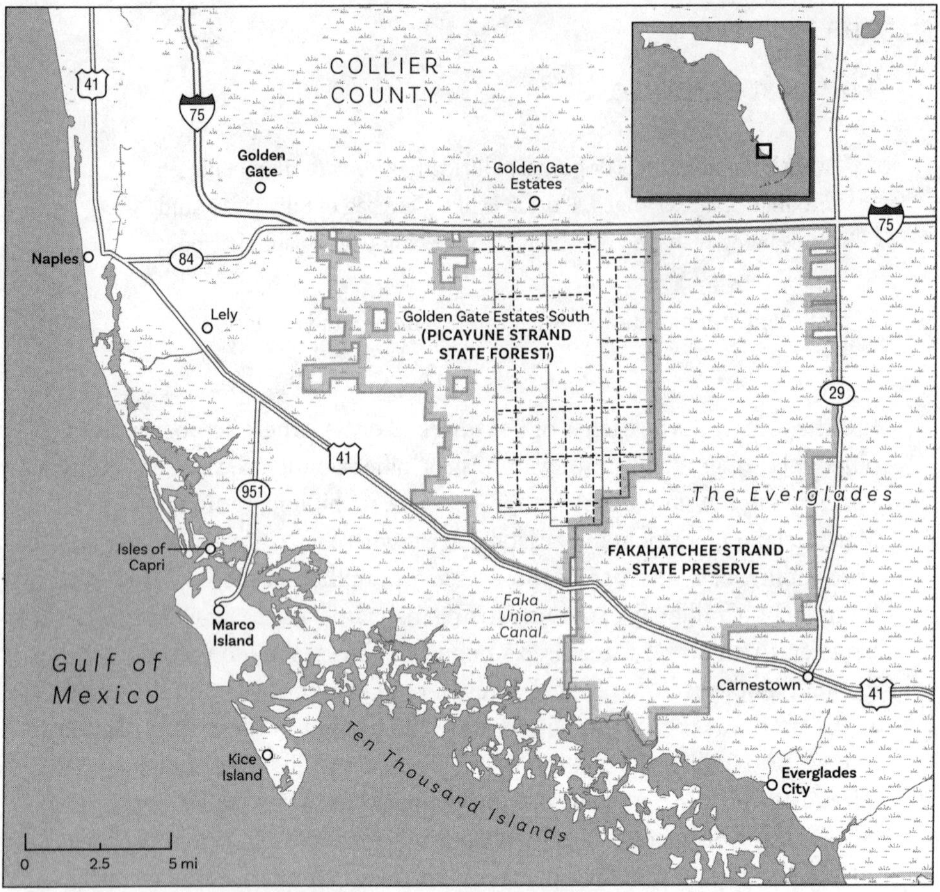


Collier County, Florida. Map created by Erin Greb Cartography.

inland; some hunting and fishing clubs as well as seasonal hotels at Naples and Everglade, which each had 200 residents; and another 400 residents in the clamming communities of Caxambas and Marco Village on Marco Island. That was it.

There were no highways, no railroads; transportation to and from the county was by boat, through tiny steamship lines such as the Everglades Transportation Company or the Marco Mail Line of Fort Myers. In fact, a map published by the Florida State Road Department in 1917 showed that the future Collier territory had just a single "hard-surfaced road," an unpaved, graded thoroughfare consisting of seashells. "It did not take us long to drive [south] out of Fort Myers," remembered a Marco Island schoolteacher. "The sand shell roadway, worn to a washboard surface, was not bad for that time, but the many 'bridges' over swampland and inlets were unbelievable. They were constructed of planks simply nailed to pilings with the cross-planks laid loose."[3] Thus, travel by car from Tampa to Miami, through Collier County, was an impossibility, so drivers went inland, to Orlando, then from Orlando to the Atlantic coast at Daytona, then from Daytona down the Dixie Highway to Miami. It was an arduous trip, which took days, so boosters on both coasts began promoting alternative routes.

The one that won, eventually, was the Tamiami Trail, a Gulf route connecting Tampa, Bradenton, Sarasota, Venice, Punta Gorda, Fort Myers, and Naples with Miami. But, at the time, building costs were the counties' concern, not the state's, so unless each of the five counties then involved in the project funded its portion of the road, the road would never get built. In 1915, Lee County approved a $175,500 bond issue to fund a hard-surfaced road to Naples, then from there a graded road to Marco, but because of the advent of World War I and local political issues, funding for the Lee portion of the Marco-to-Miami road stalled.[4] It remained stalled until mid-1923, when Florida's governor approved Collier as a county, and Barron Collier went to work. "I intend to build the Tamiami Trail through Collier County just as fast as it is possible to do it," he told a Fort Myers newspaper. "I realize that I have a great task ahead of me. But I have set my heart to that task, and am sure that [when the road is completed], many thousands . . . [will] live where only a few hundred are living [now]."[5]

Yet Collier wasn't building a road, at least not any road; he was cutting a path through the muckiest, most inhospitable portions of America's largest swamp. Therefore, to do so, he brought in some 2,000 workers, who constructed a base camp at the coastal village of Everglade, today's Everglades City, by dredging and filling a 760-acre townsite on the edge of a mangrove swamp. This was Collier's capital, his metropolis, a remote outpost near Florida's Ten Thousand Islands chain that doubled as the county seat. "No one has ever heard of Everglades [City]," wrote a *Tampa Tribune* reporter who visited the settlement in May 1925, "and it is quite a task to find anyone even who can give intelligent information as to how to reach the place."[6] It was, after all, a twenty-three-mile boat ride from Caxambas, on Marco Island, and drivers couldn't get to it by road. So, in October 1923, workers pressed north, dredging a canal and building a highway four miles into the interior, where, at an intersection called Carnestown, named for Collier's wife, Juliet Gordon Carnes, crews dredged west, to Marco, and east to the Dade County line.

The "East Trail," as it was known, was thirty-one miles long. With two huge dredgers, workers dug a canal then used the fill dirt to make a track bed for a railway, in addition to a bed for the road. But, instead of dirt or marl, they encountered solid limestone, which turned what was a difficult job into Florida's Panama Canal. For three full years, a railway car mounted with air drills drove sixteen-foot holes into the rock. There were three drills, six feet apart, so as the car moved forward and workers finished a section, blasting crews dropped forty sticks of dynamite into the center hole and twenty sticks of dynamite into the side holes before blowing them up.[7] Then dredgers scooped up the rocks, which were crushed, leveled, rolled flat, and turned into road. The work was slow, at times averaging just a hundred feet a day, not to mention dangerous. Men labored in waist-deep water full of cottonmouths and alligators, while in 1926, two workers died near Fort Myers when a lightning strike ignited 1,000 pounds of dynamite and obliterated a work shed.[8] Nevertheless, work proceeded apace through the spring of 1928, when on April 25, a caravan of some 300 cars, draped in flags and bunting and set off by a bugle, made a two-day trip from Tampa to Miami.

They stopped in Collier County, where a band of Miccosukee Indians escorted the procession to Everglades City for the first-ever "Collier County Fair and Tamiami Exposition," paid for by Collier himself. It featured a circus, a re-created Indian village with real native inhabitants, the "Royal Czecho-Slovakian Band," an aquarium with "hundreds of Collier County's finny tribe," and an assortment of "wild beasts and fowls from the fastnesses of the Everglades."[9] Collier gave a speech, as did a Native American medicine man named Assumhatchee, who said somberly: "The Seminole Indians of Florida welcome the white man's road." But "we do not know what is to become of us because the Everglades are all that we have to call home." Still, "we realize that we must give way to development" and "hope that the Indians [here] will not be entirely forgotten."[10] They were, almost immediately, when the next morning the motorcade wended its way to Miami, where it crossed the Venetian Causeway and toured Miami Beach. There, the motorcade received a twenty-five cannon salute, then, as the cars returned to Miami for a celebration at the Bayfront Park amphitheater, a pilot and his fiancé symbolically merged the two Floridas, east and west, by holding a wedding ceremony in a seaplane overhead.[11]

In 1928, Miami's downtrodden citizenry needed some good news, for as we saw in chapter 1, Florida's land boom had gone bust, terrifically; the city had been hit by a hurricane; and the soon-to-be-bankrupt Coral Gables development was $29 million in debt. The symbolic low-point, perhaps, came in February, when $1.8 million in Miami real estate securities sold at auction for just eight cents on the dollar.[12] There were no jobs, anywhere, not even in agriculture, where Florida's land developers had been so subdivision crazy that they bought up profitable orange groves and removed over a million trees.[13] This was anathema to Collier's idea of growth. "I never believe in booms," he once said. "They have a too violent effect on the real and sound values of all classes of property. The frenzied habit of selling real estate by blueprints and expecting some small investors to come along and buy lots only to sell them to some other investors has to [end] or a slump will inevitably occur."[14]

Instead, Collier favored what he called "constructive development," the building of roads, bridges, canals, railway lines, telephone lines, and other infrastructure as the basis of economic growth. Because he believed that the state of Florida should drain the Everglades to make farmland, he wanted people to plant things, "to plow down to the subsoil," as he put it, "rather than throw up a subdivision."[15] Therefore, his form of development was decidedly utilitarian: a road from Naples to Miami; a port at Everglade; a steamship line connecting cities from Tampa to Key West; a twelve-county telephone service; a hotel chain; and a bus company called "Tamiami Trail Tours." What Collier didn't build, however, were acres and acres of homes. Indeed, the first Florida land boom seems to have bypassed his county altogether except, perhaps, for two failed developments: one on Kice Island south of Marco Island, the other on Marco itself.

Laid out by a Kentucky real estate agent named Murray S. Kice, who had purchased the land from the state in 1915, "Kice-Island-on-the-Gulf" had streets named for Kice family members and 971 lots. They were platted in 1926 but abandoned when the boom ended and Kice moved to Los Angeles. (A company caretaker named Joe Dickman remained on the island, alone, for the next thirty-one years. When asked by a tourist if he, in fact, was the "Hermit of Kice Island," Dickman replied, "I've been here a long time, but I've never seen one.")[16] The county's other boom-time development was "Collier City," on Marco Island, founded in 1926 by a group of investors led by New York brewery magnates George Ehret Jr. and George E. Ruppert. Ehret and Ruppert purchased the property in large part from Captain W. D. "Bill" Collier (no relation to Barron Collier), whose father, W. T., had settled the island in 1870. They spent upwards of $600,000, staking out a 525-lot townsite with plans for roads, electric lines, sewer lines, and sidewalks, but like Kice, Collier City went bust, selling just eight lots in nine years.[17]

Had it been built, the development on Marco would have been a good one, akin perhaps to the uber-wealthy Jupiter Island community near West Palm Beach. For one, Ehret and Ruppert were incredibly well heeled—Ruppert's brother owned the New York Yankees, for example, and was a congressman, while in 1927 Ehret's father left the family a $38 million estate. Second, Marco was beautiful, a lush, mostly untouched collection of shrubland, hammock land, and even hilly areas, interspersed with bays, streams, lagoons, saltwater marshes, and mangrove swamps. It was six miles long and four miles wide, with at least five miles of white sand beach. "It is said . . . that nature has done more for [Marco] than for any other spot, island or mainland, in the entire state," wrote a person who visited the area in 1930. Its "natural scenic beauty" and its views of the Ten Thousand Islands to the south "beggar description."[18] The fishing was incredible, while Marco's 6,800 acres teemed with life. There were porpoises, manatees, panthers, bobcats, alligators, black bears, ospreys, herons, ibises, sea turtles, bald eagles, burrowing owls, and white-tailed deer. The island was also a hotbed for shell collecting and the average annual temperature was seventy-two degrees.

Yet, when Collier City died, no one noticed. Bigger and costlier projects had failed throughout Florida while the island's clam industry was booming. Between 1910 and 1946, for example, the Doxsee Clam Factory on Marco would harvest some 5 million bushels, or 300 million pounds, of hard shell clams.[19] The production was so great that in 1927, at the urging of Barron Collier, the Atlantic Coast Line extended a spur and built a railroad bridge to Marco. Unfortunately, the Depression hit. Workers overharvested the region's clam beds and jobs started to disappear by the 1930s. With no industry to speak of, Marco languished, registering fewer residents in 1950 (339) than it had had in 1920 (407). Its largest landowner was Barron Collier, but by 1933 the advertising magnate was nearly $18 million in debt. The stock market crash had killed the ad industry, so Collier did what he could to keep his empire intact. Land rich yet cash poor, he saved money by shuttering his bus company, then expanded his logging interests and leased land to cattle ranchers in Collier, before drilling for oil in the Everglades.[20]

When Collier died in New York of a heart attack in 1939, his three sons—Sam, Miles, and Barron Jr.—took possession of the holdings, but estate taxes, the bankruptcy of numerous affiliate corporations, and World War II left the family's Florida businesses in limbo. The Collier Corporation was still incredibly rich, but Sam died in a racing accident in 1950, then Miles died of polio in 1954. That meant more estate taxes, a buyout for Sam's heirs, and slow, mostly conservative investments through the 1950s. The focus was on agriculture, though a subsidiary known as the Collier Development Corporation sold residential lots in Immokalee and Everglades City and built the Pine Ridge subdivision in Naples in 1954. But Marco was a different story.

Development here required infrastructure improvements on a mass scale, not to mention a bridge, as its sole link to the mainland was a 1,500-foot, single-lane "swing" bridge that an attendant opened by hand.[21]

The Colliers had wanted to develop the island, so much so that in 1949 they cleared the property by moving the entire village of Caxambas, and even its post office, to the eastern tip of the island at Goodland Point.[22] Although the Ruppert family still owned Marco Village, some 90 percent of the island's 6,800 acres as well as most of its beaches were theirs. But what to do with it? They considered, briefly, a proposal from a Colorado man who wished to use Marco as a workers' community for an oil refinery and aluminum smelter on Kice Island.[23] But, in 1955, an ambitious Fort Lauderdale developer named Leland Loach purchased four mangrove islands to the north of Marco, about 600 acres in total, which he dredged and filled and began selling for $2,300 a lot. These were the "Isles of Capri," he called them, and though they had no drinking water or even a road to the mainland, by September 1956, people had purchased 230 of them.[24]

"Nothing seems to be phony about this," explained *Miami News* reporter Jane Wood, who visited the project in July 1956, but it is "one of the most audacious land gambles of this decade." A developer is "filling the remote mangrove swamps of the southwest coast of Florida" and turning "half-submerged patches" into dry land. Although seemingly unbelievable, she wrote, a land boom was "lapping" at Marco, a "cherished . . . almost unpeopled world of sea, sand, and sun" that "time forgot."[25] But change was coming, even to Marco, where, in 1958 a group of Miami investors opened a tiny development called "Marco Highlands" in the center of the island, and in 1959 the Air Force built a seventy-five employee missile-tracking station at its southwestern tip. Nevertheless, nothing would happen to the island, development-wise, until Barron Collier Jr. and the family-owned Collier Corporation wanted it to. They'd been waylaid by two premature deaths, but after selling some 300,000 acres to settle estate taxes in 1953 and 1959, including an extremely valuable five-mile stretch of Estero Bay between Naples and Fort Myers, by the early 1960s, they were ready to go.[26]

That's when Barron Jr. visited the Deltona headquarters in Miami. "He came to my office," said Frank Mackle in a 1992 interview, "and asked me if I'd ever been to Marco Island. He told me he thought we'd be the perfect company to develop it. . . . [So] I drove over there, and couldn't believe what I saw."[27] The white sand beach. It was mid-1962 and Frank Mackle, his two brothers, company architect Jim Vensel, and sales head Neil Bahr were standing at the north end of the island, in their suits and ties, surveying the landscape. "We were stunned," said Bahr. "Absolutely stunned. It was so magnificent. It was almost too good to be true"—the last, large, truly uninhabited waterfront stretch of southern Florida that you could build on, and the Collier family was handing it to them on a plate.[28] The terms were amazing: in a deal finalized in March 1964, the Deltona Corporation and an investment group made up, most notably, of Barron Collier Jr., sister-in-law Isabel Collier Read, the Gerry Brothers, and Canadian media magnate Peter Thomson split ownership of the island and several nearby islands and property on the mainland, in total some 10,000 acres, 50–50 through a joint company called the Marco Island Development Corporation (MIDC).

In exchange, the Colliers received approximately $7.5 million, paid for by Deltona and the other non-Collier group members, while Deltona and the group, Colliers included, provided $4.5 million more in capital.[29] However, the Colliers wanted the Mackles so badly, apparently, that they allowed Deltona to pay its share of the $7.5 million at zero percent interest, with payments beginning in 1968 when lot and home sales had already begun. Not only that, the Colliers permitted Deltona to cover its portion of the $4.5 million in capital by using a bank loan with Marco Island itself as collateral, then gave the company 5 percent of construction costs and 3 percent of lot sales. In other words, the Colliers had handed the Mackles half of Marco and the right to develop it for no money down. "It seemed as if the Colliers were . . . giving their island away," wrote Douglas Waitley in his 1993 book *The Last Paradise: The Building of Marco Island*, but as one official put it, the Mackles were "the foremost developers in Florida, if not the nation." They "had the planners, the engineers, the equipment, and the marketing organization" to make the project a success.[30]

That was easier said than done, because Marco was remote. It needed a bridge, for one, and not just a swing bridge but a quarter-mile-long concrete structure with a fifty-five-foot-high span. It then needed roads, ditches, canals, seawalls, bulkheads, a wastewater treatment plant, an electric plant, and a water purification system for water it didn't have. The closest supply, in fact, was a lake owned by the Colliers, on the mainland, which required nine miles of pipe.[31] But the Mackles were insistent. They understood how hard it would be to build Marco, but they had the know-how and experience and also knew the rewards: a reported 10,839 homesites priced at $2,495 and up for "inside" lots and $5,495 and up for waterfront and golf course lots. Then, they would offer homes in twenty-two different models, ranging from the one-bedroom, one-bath "Granada" to the three-bedroom, four-bath "Martinique" priced at $14,900 to $41,500 apiece. But, since Marco was a

In the mid-1960s, the Mackle brothers' Deltona Corporation teamed with Florida's Collier family to develop Marco Island, a near-pristine barrier island south of Naples near the Everglades, and turn it into a planned community of some 8,000 waterfront lots. Courtesy of Marco Island Historical Society.

planned community reminiscent of Port Charlotte, only pricier, architect Jim Vensel set aside 10 percent of its land area for parks, churches, hospitals, government buildings, and schools.[32]

What was different about Marco, however, and what set it apart from Port Charlotte or Key Biscayne, for example, the Mackles' other island development, was Vensel's commitment to "finger islanding": replacing its environmentally sensitive wetlands and mangroves with row after row of seawalls—with residential peninsulas, divided by canals, buttressed by seawalls. It was an island of seawalls, 180 miles of them, on a land mass just a tad smaller than Manhattan. In all, Vensel planned a community in which 8,000 homesites were waterfront, a seemingly impossible number, requiring the dredging and filling of 35 million cubic yards, or roughly 77 billion pounds, of dirt.[33] It would come, for the most part, from the canals workers that were digging from the bottom of two bays and the waters surrounding a key, and from the floor of the Big Marco River, scooped up by excavators pulling buckets, called draglines, or sucked to the surface using pumps. Although an amazing amount of work, the dredging of fill dirt from bays and rivers was preferable to buying dirt from the mainland—at upwards of fifty cents a yard—because dredged dirt from the state cost a nickel.

In a process repeated again and again throughout Florida, developers would buy a mangrove island or a low-lying stretch of coastline or beach and ask permission from the county to establish a bulkhead, a boundary-forming seawall at the edge of the property facing the water. Counties had different policies but typically set "bulkhead lines" far out into the water and gave permissions as a matter of course. Then, developers went to the state, which, by law, owned all land under navigable water, from river and lake beds and bay bottoms to even the seafloor stretching three miles into the Atlantic and ten miles into the Gulf. The state administered these "submerged lands" through an agency known as the Internal Improvement Fund (IIF), which, when petitioned, allowed developers to, as one journalist put it, "dredge from the public bottom outside the bulkhead line all the sand that's needed to fill the submerged land inside the bulkhead line."[34] The IIF also leased submerged lands to oil companies and even sold parcels outright to developers of coastal property wishing to extend bulkhead lines sometimes thousands of feet into the water. The only caveat was that the Army Corps of Engineers, a federal agency, had to approve the developer's plan.

Dating to 1775, when the Second Continental Congress created a three-person engineering unit to build fortifications at Bunker Hill, the Corps had grown by the mid-twentieth century into an infrastructure-building behe-

In a process known as "finger islanding," developers would create waterfront property by dredging and filling critically sensitive wetlands and replacing them with dry peninsulas separated by canals and buttressed by seawalls. Courtesy of the Florida Photographic Collection, State Archives of Florida.

moth, a 35,000-person bureaucracy within the military whose workers were civilian. In addition to defense projects, the Corps had as its mission, through the General Survey and Rivers and Harbors Acts of 1824, the overall maintenance of America's navigable waterways. That led to flood control, the building of dams and locks, levees, canals, aqueducts, reservoirs, hydroelectric plants, water plants, outdoor recreation areas, artificial lakes, and even beaches—in all, some 4,300 active civic works projects in the 1960s at a cost of $32 billion.[35] The Corps was adept at dredge and fill, having straightened, cleared, opened, or deepened virtually every navigable thoroughfare in the United States. In the 1880s, for example, it built the East and West Potomac Parks near the Mall in Washington, D.C., using 12 million cubic yards of fill dirt from the Potomac River. It dredged the Houston Ship Channel in the 1920s, the Gulf Outlet Canal in New Orleans in the 1950s, and once had fourteen different hopper-dredgers at work in New York Harbor alone.[36]

And since its primary mission was to keep waterways clear, free of sandbars and silt, the Corps dredged, it seemed, in perpetuity. Shallow bays were bad; shallow rivers were bad, so if the Mackles wished to pull fill dirt from the waters near Marco, the Corps absolutely approved. In fact, when developers asked the Corps for dredge and fill permits the answer, invariably, was yes. There were no public hearings: only an article in newspapers in which the district office in Jacksonville announced that the Bomford Investment Co. of Tampa, for example, wished to enclose a beach with a seawall and dredge an eight-foot-deep, 74-foot-wide boat channel 1,500 feet into the water. Anyone opposing the plan could write a letter to the "District Engineer, P.O. Box 4970, Jacksonville," but objections based only on navigation—and not land use or environmental issues—would be considered.[37] Similarly, the Mackles' Marco Island announcement appeared in the Fort Myers *News-Press* on 12 July 1964. In it, the Corps gave a six-week deadline for written protests, then approved the dredge and fill permit that October.[38]

However, the Corps' dredge and fill permits were for three years, while the Mackles had hoped to finish the project in fifteen. Therefore, they divided the job into five parts: (1) the Big Marco River area to the north of the island facing the mainland, to be permitted first; (2) the Roberts Bay area to the south of the island facing the ocean, to be permitted in 1967; (3) the Collier Bay area to the northwest, set for 1971; and (4) the Barfield Bay area to the southeast and (5) the Big Key area to the northeast, both set for the mid-to-late 1970s. While this seems daunting, one author explained that getting Corps permits "was such a routine matter" that the Mackles began dredging and filling the Big Marco River area *before* its permit was approved.[39] They more or less had to, because the island's new water supply, brought in from the mainland through pipes under the Big Marco River, an electric plant, two streets of model homes, a hotel, and a welcome center were slated to open on 31 January 1965. These and other improvements cost the Mackles some $4 million, the bulk of their start-up money, so it was imperative that MIDC begin selling lots as quickly as possible.

While Marco was a ritzier development than, for instance, Deltona, its economic model was the same: big ad budgets, nationwide sales dinners, bus tours, plane tours, celebrity "spokespitches," and sales—installment sales—of yet-to-be-dredged and yet-to-be-filled lots, not just in the Big Marco River area, which had been permitted, but in all five areas at once. They called it, like they did at Deltona, "coordinated growth," with cash sales in the first area, three years of payments in the second area, and so on. But whereas the cheapest homesites in Deltona were $995, with payments of $15 a month, the cheapest in Marco were $2,495 at $30 a month. And those were inland sites. The more numerous canal-front sites were even steeper: a minimum $5,495 at $65 a month, with several high-end oceanfront sites listed for $19,000. Marco "isn't a low-cost retirement community for the factory foreman or the policeman," wrote one northern newspaper. "It's aimed at the family of the professional man or the executive who decides to retire in style, but quietly. It's also aimed at the folks with money who want 'a place in Florida' to come to during the winter."[40] With an air of exclusivity but with installment plans and ad campaigns geared to the upper-middle-class masses, Marco attracted buyers in droves: 25,000 people toured the island on opening day.

They were directed to MIDC's welcome center by twelve Collier County sheriff's deputies, then herded onto tour buses (and even a boat) for carefully staged walk-throughs at each of the development's model homes, where the atmosphere, exclaimed one agent, was "electric."[41] Within six weeks, people had purchased over a thousand homesites, and by August 1966, just eighteen months into the project, MIDC had sold $3.5 million in homes and apartments and $22 million in land. It had built, or was in the process of building, 125 homes, a grocery store, a hardware store, a post office, a service station, a 100-room hotel, a 48-unit condominium, a 78-unit apartment complex, a country club with an 18-hole, professional-grade golf course, a yacht club, two restaurants, six miles of seawalls, twenty miles of utility mains, and eighteen miles of roads. Then, in 1967, Marco got a boost from the state, and the county, when officials approved a $2.1 million bond issue for a much-needed bridge across the Big Marco River (connecting the development to the Isles of Capri and the mainland, the bridge cut the distance to Naples to just eighteen miles).[42] "In the brief [period] since Marco Island opened," said Frank Mackle, "we've made tremendous strides. In all respects, we're running well ahead of schedule. . . . [Things are] taking shape so fast, it's like looking at tomorrow."[43]

OF COURSE, "TOMORROW" INCLUDED an Army Corps of Engineers permit for area two, the Roberts Bay area, which the Mackles applied for in September 1967. They expected smooth sailing, but in the three years since their first permit, the Corps' long-standing regulatory emphasis had changed. Now, instead of a singular focus on navigable waterways, the Corps would also consider the environment, deciding whether or not a given dredge and fill project would "unreasonably impair natural resources . . . including the fish and wildlife and recreational values thereof."[44] The Corps hadn't wanted to approve or deny projects based on environmental factors, because its job, traditionally, was to transform nature, to master it, to act, as one person described it, as a tireless beaver "whose instinct for dredging and filling know[s] no bounds."[45] But by the 1960s, the world's burgeoning environmental movement had taken root in America, beginning, many believe, with the 1962 publication of *Silent Spring*, a book by scientist and author Rachel Carson, who described the deleterious effects, both to plants and animals and to humans, of commonly used pesticides such as DDT.

Then, in quick succession, came the Clean Air Act (1963), the Water Quality Act (1965), the Endangered Species Preservation Act (1966), and most notably for the Corps, a little-known but hugely important estuaries-preservation bill introduced by Rep. John D. Dingell of Michigan (1967). A hunter and fisherman in his own right, Dingell was chair of the House Subcommittee on Fisheries and Wildlife Conservation, which in March 1967 held hearings on America's estuaries, the environmentally sensitive fish breeding grounds along coastlines where freshwater and saltwater mix. Dubbed "nurseries of the sea," estuaries are found in semi-enclosed bodies of water, such as bays or lagoons, and provide spawning and/or nursery grounds, or general habitat, to some 80 percent of fish and shellfish species and 85 percent of waterfowl and migratory bird species on earth. As Assistant Secretary of the Interior Stanley A. Cain would testify, the United States had nearly 8 million acres of "important estuarine habitat," but because of dredging and filling by land developers and agencies such as the Corps, it had lost in two decades a gut-wrenching 568,800 acres of it, or 7.1 percent.[46] That was an area almost the size of Rhode Island.

Nevertheless, the best way to mitigate damage, insisted Cain, was the so-called Dingell bill, which, in addition to a Corps permit, required anyone wishing to dredge, fill, or excavate any portion of an estuary anywhere in the United States to seek permission from the Secretary of the Interior, who, in turn, could deny requests that would "unreasonably impair" wildlife, water quality, sport and commercial fishing, recreation, or aesthetics. However, the Corps feared that if the bill passed, it would give the Department of the Interior too much say over its own projects, which were not just damaging environmentally but the very definition of pork. As the *Washington Post* put it, the Corps is "like a congressional preserve" and "its civil works budget consists almost entirely of earmarks requested by individual members of Congress and endorsed by the Corps." So America "doesn't really have a water resources policy; just a pork-barrel water resources agency" that "builds pet projects in congressional districts across the country."[47]

As a result, the Corps was "untouchable in Congress," a sacred cow, which meant that though well-intentioned, Dingell's two-permit proposal for approving dredge and fill operations was moot. However, in July 1967, as a bone to critics, Secretary of the Army Stanley Resor and Secretary of the Interior Stewart Udall signed a memorandum of understanding in which, prior to permitting a project, the Corps' district engineers would "coordinate with" and seek "technical advice" from directors at Interior. The Corps would have final say over who dredged and filled, but Interior would review applications to determine whether or not a given dredge and fill project was overly damaging to the environment. If it was, the Corps would deny the permit outright or, the memorandum stated, "encourage the applicant to take steps that would resolve the objections to his work."[48] That seemed easy enough, but before anything changed, the Corps' newfound environmentalism had to stand up in court to a pair of dredge and fill land developers in St. Petersburg who were trying to add an eleven-acre peninsula to a mobile home park on Boca Ciega Bay.

Located to the west of St. Petersburg at the north end of the mouth of Tampa Bay between the Pinellas Peninsula and St. Petersburg Beach, Boca Ciega Bay was a shallow estuary, fed by estuaries, some 20,000 acres in size. Its beaches, islands, bayous, and wetlands provided a critical habitat to numerous endangered species, including West Indian manatees, wood storks, red-cockaded woodpeckers, and Kemp's ridley sea turtles, which today are the most endangered sea turtles in existence. But, by 1968, Boca Ciega Bay had lost some 15,000 acres to dredging, while almost 5,000 acres had been filled.[49] It had four causeways, several man-made isles and peninsulas, and dozens of finger islands; the bay was filled with them, and each housed subdivisions like Yacht Club Estates, Belle Vista, and Vina Del Mar. "Pushed out into its once broad and sparkling waters," wrote one journalist, these "finger fills" have reduced the bay to "a series of narrow boat channels."[50] They also devastated sea life, as dredged bay bottoms were often too deep for sea grasses, in particular the *Halodule wrightii*, or shoalgrass, where much of the bay's fish and shellfish species lived.

What is more, dredge and fill operations removed sand from the bay floor, leaving in its place a lifeless substrate of clay, covered in silt, where nothing grew. In fact, one 1966 study by the U.S. Bureau of Commercial Fisheries found that in northern Boca Ciega Bay, where dredging had been prolific, 85 percent of bottom soils were silts and clays. Forming a mucky, slimy, polluted layer some *thirteen feet deep*, the Boca Ciego Bay floor contained no large forms of life, stated the bureau, "and low numbers of only two small worms."[51] Now add 9.1 billion gallons of sewage effluent that wastewater treatment plants dumped into the bay every year and you had a dead sea, where people paid top dollar for canal-front lots that had been dredged and filled. As biologist Ken Woodburn put it in 1964, "Dredging and filling is the most profitable [method] for creating waterfront real estate. Cities are built at the water's edge on land which has been pumped from the estuary." But cities need "connecting fills." They need causeways and bridges, which, in turn, disrupt the bay's "normal circulatory regime." At "the present rate," the eating up of "marshes and intertidal plains so necessary for [marine life] . . . will destroy these areas as a resource."[52]

Developer Lee Ratner's Bayway Isles project in St. Petersburg was a case in point. In 1956, Ratner, who also owned the Lehigh Acres Development Corporation, optioned 497.66 acres of submerged land at the mouth of Boca Ciega Bay between what became Florida Presbyterian College (now Eckerd College) and St. Petersburg Beach. His plan was to enclose the bay, more or less, by building two large islands for a high-end homesite and condominium community connected to the mainland. Never mind that Ratner's design would turn what had been a two-mile-wide tidal passage into just three channels 1,200 feet, 500 feet, and 400 feet wide. Yet somehow, and in spite of its mission to keep waterways clear, in 1957 the Corps of Engineers upheld a permit for the project, as did local authorities, and Ratner began turning the bay into a lake. By early 1959, he had dredged some 9,000 yards of fill dirt, but stopped when a local organization called the Alliance for Conservation of Natural Resources and several other groups questioned his permit in court. The alliance had first sued Ratner in 1957; it lost, then appealed, then lost again, before taking the case to the Florida Supreme Court, which, in 1960, essentially sided with Ratner by refusing to rule. He could build the isles if he wanted to, and by 1965 he was selling homes.[53]

Meanwhile, other Boca Ciega developers continued to dredge and fill, most notably Alfred Zabel and David H. Russell in the bayfront community of South Pasadena. Incorporated in 1955 on just 465 acres, including some submerged acres, South Pasadena was little more than a trailer park at the base of a causeway. The city (and its 150 residents) needed land to expand, so it dredged and filled to the edge of its territory, then annexed a neighboring fill in Boca Ciega Bay called Pasadena Isles. While it couldn't issue its own permits, South Pasadena supported any project that added to its tax rolls, including one in 1958 for an eleven-acre, catcher's-mitt-shaped extension to Zabel and Russell's Causeway Village trailer park.[54]

Although not as lucrative as the high-end development of Bayway Isles, the Causeway Village trailer park made perhaps $50 a month on each of its 300 or so lots. With some ten to fifteen lots squeezed into an acre, an eleven-acre extension made dredging and filling and building a seawall worthwhile. Therefore, in 1958, Zabel and Russell applied for a "fill-and-bulkhead-line" permit from the Pinellas County Water and Navigation Control Authority. But anger over the "Ratner fill" and dozens of fills like it had local residents on edge. They hired an attorney and told the authority in four different hearings that dredging and filling were bad for wildlife, that they disrupted tides, lowered property values, affected people's recreational activities, and silted the bay bottom. "There used to be a long white beach across the bay where residents went swimming," commented a man who lived near the proposed fill. "After the dredgings, no one went swimming anymore."[55]

While not exactly defendants, in a legal sense, Zabel and Russell asked General Robert G. Lovett, a former district engineer with the Corps, to serve as a rebuttal witness. Lovett worked for Rader and Associates, a Miami firm that had helped plan both the Cape Coral and Lehigh Acres developments, and he would testify that yes, Boca Ciega's fill projects had been bad for the environment, but there'd been so many projects that "the effect of this fill would be nil." "You mean so much damage has been done [to the bay] that this won't make any difference?" asked an attorney for the protestors, incredulously. To which Lovett replied, "Yes, sir."[56] Although refreshingly blunt, Lovett's argument failed to convince the authority, which in 1959 refused to permit the project because of its "adverse material effects" on the environment. However, Zabel and Russell appealed, insisting the burden of proof was with the county, not them, and that because the county had failed to conduct its own studies, let alone publish them, it had in effect taken their submerged property without paying for it. The case went to the state supreme court, which, in 1965, in a controversial decision, sided with Zabel and Russell.[57]

Next up for the developers was formal approval from the state's IIF board, then, of course, from the Corps. In 1966, with little fuss, the board ruled in the affirmative, but the Corps took the unusual step of issuing a call for objections based on factors "other than navigation." It also extended the deadline, then agreed to host a public hearing in November 1966.[58] Held in St. Petersburg in the recently opened Bayfront Center—which, ironically enough, had been built on submerged lands using fill dirt—the hearing drew at least a hundred people, mostly residents and members of conservationist groups such as the Save Our Bays Association and the Audubon Society. Through the summer of 1966, these groups had barraged the Corps with thousands of form letters denouncing the fill and teamed with Pinellas authorities to host twenty-eight different witnesses, including an engineer, a biologist, a geologist, a sanitation expert, a state senator, and a representative from the U.S. Fish and Wildlife Service, who together spoke for nearly five hours.

By contrast, those for the fill were in the minority, a total of two speakers, and included Thomas M. Harris, the attorney who had represented the developers in court. To boos and hisses from the crowd, Harris admitted that yes, he understood why people were objecting "from a conservation standpoint," but insisted, "It's our land. We can do with it what we want."[59] That was the crux of the matter, and the Corps, heretofore, would have almost always agreed with him. But in March 1967, less than week after the House Subcommittee on Fisheries and Wildlife Conservation had held hearings on the Dingell bill, the Corps found its environmental conscience and nixed the permit altogether. "As far as I know," said a shocked biologist at the U.S. Bureau of Commercial Fisheries' St. Petersburg office, "this is the first time the Corps has denied a permit based upon conservation. . . . [This is a] monumental decision."[60]

Although appeals and further appeals would follow, developers took note: The Corps wouldn't be as malleable, or as mindlessly pro-development, or as quick to approve dredge and fill permits as it was before, and under first-year governor Claude Kirk, neither would Florida. By 1967, the state was literally awash in shit. Its air and water were polluted; its wetlands were dwindling; and it was feeling the effects of overpopulation. As the Associated Press put it: "All over Florida, waters once clear, blue, and sparkling now are a dirty brown, yellow, or black. Cities and towns are defiling lakes and streams with human sewage," the state's "industrial plants have turned rivers into receptacles for toxic waste," and "real estate developers cloud the waters with silt." Now, "the time has come, many feel, that Florida must decide whether it will take bold steps to control the wastes of industry, agriculture, and its growing population or strangle in them."[61]

Warning signs were everywhere, and not just in Boca Ciega Bay or in urban areas, but in sparsely populated regions of southern Florida, where sprawling land giant communities mixed awkwardly with small-town infrastructures and where pollutive industries such as sugar growing or phosphate mining were key. Twice in the 1960s, for example, thousands of gallons of leaked phosphate sludge from a pond in Fort Meade coated the Peace River with "a stream of flowing pancake batter" that killed all of the fish. Along three-quarters of its 102 miles, "clay-like deposits [clung] to the river bed," turning the water yellow and sending countless fish carcasses as far south as Port Charlotte.[62] In 1966, in nearby Fort Myers, where each day raw, untreated sewage flowed into the Caloosahatchee River from residential areas at Russell Park and Colonial and McGregor, the State Board of Health found shellfish beds downstream at Punta Rassa ridden with coliform and thus too dangerous to eat. In both Charlotte and Lee counties, state bans on shellfish harvesting were common—the water was that polluted—but in 1973, the state issued a ban for all of Southwest Florida, from Manatee County in the north to Collier County in the south, when three people suffered "paralytic symptoms" after eating clams harvested from Sarasota Bay and Siesta Key.[63]

Meanwhile, in LaBelle, the rural cattle community where the General Development Corporation (GDC) was planning Port LaBelle, the lack of a wastewater treatment plant meant that 60,000 gallons of urine and excrement flowed into the Caloosahatchee every day. Trapped between the W. P. Franklin lock at Olga and the Moore Haven lock, it eventually flowed west, to where the city of Fort Myers tapped the river for its drinking water.[64] But that wasn't the worst of it. For decades, the dredging of deep navigational channels, not to mention labyrinthine canal networks, throughout the Caloosahatchee had allowed saltwater from the Gulf to seep upstream and into the water supply. The reason is that saltwater is heavier than freshwater; therefore, in southern Florida, east and west, it stays near the coast in a kind of brackish equilibrium as long as rainwater flowing from the center of the state flows outward at a higher rate of pressure than the seawater flowing in. But drought, population growth, and heavy irrigation by the sugar and citrus

A typical canal view, Marco Island. With a driveway in the front and waterfront access in the back, these houses were the epitome of Florida living. Note the seawall construction in the foreground. Courtesy of Marco Island Historical Society.

industries, among others, reduced the amount of freshwater coming out of rivers, which, at the same time, were now deeper as a result of dredging.

"What worries us in planning for future water supplies," lamented a state geologist in 1969, "is that rivers are being extended inland by digging canals to establish waterfront property. That brings creeping saltwater."[65] The problem was especially acute in Cape Coral, which had more canals than any other city in the world. There, Gulf American built a water treatment plant, but only after it had dug wells for some 8,000 homes near the Caloosahatchee. The wells were cased in metal, so as saltwater intruded through canals and then moved laterally, underground, it ate away at the casings and worked its way into the upper Hawthorn aquifer, where the city got its drinking water. By 1981, officials with Cape Coral's "Saltwater Intrusion Office" had detected and plugged some 800 wells with chloride levels higher than 500 milligrams per liter, when 250 or higher exceeded state limits for drinking water, and anything in the 625 to 675 range was harmful to plant life. "Time and nature will take [a] toll on steel casings," said one official, and if too much chloride gets into the aquifer, "you might as well pack your bags and move away."[66]

Either that or buy water at high cost from other municipalities or build wildly expensive reverse osmosis (RO) water treatment plants that took brackish water (or even seawater) and made it palatable to drink. Although RO plants were costly to build, by the late 1970s, Florida had at least seventy of them, producing some 25 million gallons a day in sixteen coastal counties.[67] At $10 million to $15 million apiece, they consumed electricity at an enormous rate by blasting untreated water at 445 pounds per square inch through dozens of cylindrical permeators that grabbed salt molecules and other impurities to produce drinking water. The process wasn't cheap. In 1983, for example, an RO plant in Sarasota drawing from a brackish aquifer spent, on average, seventy-one cents to produce 1,000 gallons of drinking water. By contrast, a plant in Boca Raton drawing from a freshwater aquifer spent just thirty cents per 1,000—a difference and a savings of 58 percent. To be sure, RO was an amazing technology, and for places like Cape Coral a miraculous technology, but as John DeGrove, a member of the Florida House Speaker's Task Force on Water Issues, put it: "It's infinitely cheaper to protect groundwater" than it is trying "to figure out exotic and expensive ways to treat it after you've ruined it."[68]

What Florida needed, of course, was to improve its environmental standards across the board. Sewage, water quality, air pollution, and wetlands preservation measures all needed tightening, but no one in power had the will or the stomach to do it—that is, until Claude Kirk arrived. The maverick Florida governor was a pro-business Republican, ostensibly, but had a conservationist streak and claimed as a campaign adviser the millionaire environmentalist Nathaniel P. Reed. Tall and thin with a passion for green causes, the thirty-four-year-old Reed grew up in Greenwich, Connecticut, but spent winters on a barrier island near Hobe Sound, Florida, where his heiress mother Permelia Pryor Reed and theater-patron father Joseph Verner Reed developed the property into a low-key haven for America's super elite, such as the Mellons. After college and a stint in the military, Reed moved permanently to Florida in 1960, where he worked at the family's Jupiter Island Club and managed a citrus ranch. "I guess I became seriously interested in natural resources," remembered Reed, "when I took over the management of the ranch. In short order, I learned about flood control, game and fish, water management, pollution . . . the whole picture."[69]

While doing business in and traveling throughout the state, Reed was shocked by what he saw: devastation. "Man was remaking my Florida with the heaviest of hands," he wrote in his 2016 memoir. "Development at any cost was the goal," and "low taxes, combined with the advent of air conditioning and cheap airfare, [had] encouraged a Florida real estate boom unrivaled since the 1920s. . . . Well-financed developers who often were aided by Florida's politicians . . . clear[ed] vast mangrove jungles and fill[ed] life-sustaining swamps to make room for roads, homes, and waterfront property. Thousands of freshwater wetlands were being drained" and "productive farmlands turned into gated golf course communities." But, those who opposed development, stated Reed, were dubbed "strange people in tennis shoes" who couldn't "understand the desire to transform Florida into a new playground inhabited by everyone from the ultra-wealthy to those who could barely subsist on Social Security."[70] Something had to be done, and Reed, the millionaire activist, was the man to do it.

He first met Kirk in 1965, joining the campaign as his environmental adviser and writing a sixty-two-page conservation "white paper" for the candidate in 1966. In it, Reed discussed everything from water hyacinth removal and pollution standards to the purchase of Golden Gate Estates, which Reed insisted the state should buy from Gulf American before the Rosens sold it off. When Kirk won the governorship and moved to Tallahassee in 1967, Reed moved too, working for a dollar a year as an assistant to Kirk in charge of conservation. Given "an office, a secretary, and not much more," joked Reed, his first task was to assist Kirk with tightening restrictions on the sale of submerged lands. That, for Kirk, was the easiest fix, by far, as dredging and filling had riled voters from both parties while the IIF had the annoying, even unethical habit of under-appraising state properties then using the profits as a kind of extra-legislative slush fund for things that had nothing to do with the environment. From 1961 to 1965, for example, the IIF sold some 7,000 acres of submerged property for the appallingly low price of $287 an acre. It then used the money for building livestock show pavilions at county fairgrounds, remodeling the governor's office and the Florida Supreme Court's law library, and planning additions to the new Florida Atlantic University in Boca Raton.[71]

Therefore, in January 1967, the newly installed Kirk, who, by law, belonged to the five-man IIF, joined fellow member and Democratic Attorney General Earl Faircloth and two others in declaring a moratorium on the sale of submerged lands. The goal in declaring the moratorium was to give legislators time to consider a bill establishing "sound, clear, consistent, and comprehensive policies" regarding the sale of submerged properties, which Kirk (through Reed as his ghostwriter) claimed needed "maximum protection." With overwhelming support from both the house and the senate, Kirk signed Florida's first-ever submerged lands bill into law in July 1967. The bill was a "conservation triumph," wrote observers, and required the IIF to conduct environmental impact studies before selling submerged lands. If the studies showed that a given project was "contrary to the public interest," then that project was denied.[72] Then, in 1969, as a further step, the IIF began pressuring counties into setting their bulkhead lines at the mean high water mark, rather than on a case-by-case basis in which some dredged developments stretched hundreds of feet into the water.

When, by mid-1969, not a single coastal county had bothered to comply with the directive—insisting a uniform fill line was bad for business—the IIF voted to suspend dredge and fill permits while declaring yet another moratorium, and waterfront development stalled.[73] It stalled, most notably, in Sarasota, where the land giant Arvida Corporation, so named for the late Alcoa chairman and Florida property tycoon Arthur Vining Davis, planned to build a $65 million hotel, marina, and residential community on 116 submerged acres near Lido Key.[74] The moratorium would have affected Marco, but the Corps of Engineers had already been sitting on MIDC's Roberts Bay permit for nearly two full years. It had been waiting, it seems, for the Department of Interior's Fish and Wildlife Service to study the project and issue a recommendation, which, when it finally came, was a no. The project was destructive, it said. It killed mangroves, and hurt fishing, and polluted the water, and had a terrible effect on the environment.

Yet, try as they might, officials at Fish and Wildlife couldn't stop the permit; only the Corps could. So the Mackles pressured the agency by contacting Key Biscayne resident and Florida Senator George Smathers, as well as former Florida governor, now senator, and Senate Appropriations Chair Spessard Holland. The Mackles knew both politicians personally, especially Smathers, a friend and confidant who golfed at Marco and who was adept at delivering pork. In 1969, as a favor to the Mackles, both Smathers and Holland reached out to Undersecretary of the Interior Russell Train, who steadfastly upheld the recommendation, then to Brigadier General F. P. Koisch, an influential civil works director within the Corps. Apparently, Koisch got the message, overruling Train and recommending to Secretary of the Army Resor that the permit be approved.[75] However, Resor, who, as the U.S. Army's chief civilian administrator oversaw the troop buildup in Vietnam, acted cautiously, for in February, a district court in Tampa had issued the first of three rulings in the Zabel and Russell case, finding that the Corps of Engineers had overstepped its authority in denying the men a permit for reasons other than navigation.

The Corps would appeal—and win—but, in December 1969, Resor had no choice but to approve the Roberts Bay permit, though to mitigate environmental damage, he required the Mackles to perform a dozen or so "ecologically-oriented services," which *Forbes* quipped "read like a shopping list from the Sierra Club."[76] Among other things, they had to set aside some 102 acres as a wildlife sanctuary, construct a man-made mangrove island in the center of Roberts Bay "as an experiment in environmental research," and hire an ecologist to assist with planning. Then Resor, in ominous declarations in letters to both Claude Kirk and the Mackles, warned that in the future, developers should avoid selling coastal properties before the permits were approved and that the granting of the Roberts Bay permit did not "necessarily mean that future applications for work in the same area [would] be similarly granted."[77]

The message was clear, but either out of ignorance or arrogance, or a belief among the Mackles that Marco's three remaining permits were, as Frank III put it, "a routine—if more involved—process," the company continued to sell. The brothers were bullish even, and soon bought some 7,100 additional acres on the mainland across from Marco, including the so-called Collier-Read Tract, despite the fact that in December 1968 they'd taken control of MIDC by giving the Colliers and other investors 136,250 shares or some $6.8 million worth of Deltona stock. The island, therefore, was theirs, and by 1970 they would sell some 77 percent of it, including 5,325 lots worth $61 million in areas that hadn't been permitted yet.[78] That was a risky move because lots sold in 1965 were due for delivery in 1973. So the Mackles began pushing state and federal authorities to permit the island's three remaining areas as quickly as possible. To that end, they hired the former head of the Republican National Committee as a lobbyist, marshaled environmental evidence—even claiming at one point that Marco's nearly lifeless canal network would generate more phytoplankton than the mangroves they were replacing—and struck a deal with the state to extend dredge and fill permits that had lapsed.[79]

Signed in October 1971 between MIDC and the IIF, the deal was as follows: The state would extend dredge and fill permits for Roberts Bay and the three remaining areas if the company deeded to the state some 4,000 acres near Marco, including several sizable keys and two and a half miles of beachfront on Kice Island. Although one official called it "the fairest, most realistic way to settle a bad situation," the MIDC-IIF deal had nothing to do with the Corps, which, with the passage of the Federal Water Pollution Control Act Amendments of 1972, now had even greater authority over dredge and fill.[80] For while its jurisdiction had traditionally extended to the mean high water mark, it now encompassed the "waters of the United States," meaning all of them, everywhere, from rivers, lakes, and seas to swamps, fens, mudflats, sandflats, prairie potholes, and bogs. That meant that for MIDC, the Corps could deny a permit for damage it would cause not just to bay bottoms but to a creek or a slough more than a mile from the Gulf. Of course, developers resisted the amendments in court, claiming "waters of the United States," as a regulatory term, was too broad, but rulings in 1974 and 1975 put the Marco permits in doubt.

But, by that point, Deltona was in doubt, the victim of a collapsing land sales industry that had bankrupted not one but two Florida giants: Cavanagh Communities, which owned the circular Rotonda development near El Jobean, and GAC, the Gulf American successor that had been the largest corporation (of any kind) in the state. The reasons were complex but stemmed, in part, from the 1973 OPEC oil embargo, which quadrupled petroleum prices, inflated transportation and building costs, and made everything more expensive. What is more, the prime rate, the annual percentage rate financial institutions charged Deltona and other major companies for loans, jumped above 7 percent between 1972 and 1974, which killed earnings, as did double-digit inflation rates that eroded the value of installment contracts and made it harder for people on fixed incomes to buy homes. The bottom line? The installment land sales model that had served the industry so well in the past—that is, the buying of huge tracts of land, cheap, then dicing them up and passing improvement costs onto the buyers—was, if not wholly unprofitable, much trickier to manage. As Arvida president John W. Temple stated: "It was a fad business and the fad passed."[81]

Big tracts in desirable locations were scarce; they cost more and took more to develop, while investments made at the front end of a planning cycle were harder to recoup at the back end. As S. Hayward Wills, the outgoing chairman of GAC put it, "I don't disagree with the philosophy that the little man is entitled to buy Florida land, but I don't see it as a long-term, realistic approach. Rather, I expect that in the future, the business will evolve into selling only . . . developed property. The present way has to end because there's too much upfront expense."[82] Thus, ailing financially, land giants diversified: GDC into public utilities delivery and Deltona into lumber production, property management, and the building of prefab storefronts for customers in

Iran.[83] But, with two-thirds of its income coming from land sales and nearly all of that from installments, Deltona's stock price fell from over $70 a share in 1969 to a paltry $3 a share in 1974. Even more amazing, in the final quarter of 1974, Deltona, the once-great builder of entire Florida communities, sold just thirty-seven homes. It was $145 million in debt, and as the Corps held hearings on Marco in September 1975, the company faced some $61 million in refunds if the project went belly up.[84]

The Mackles had demonstrated "a monumental breach of good business judgment," explained Hal Scott, president of the Florida Audubon Society to a packed house of over 1,000 pro-development Marco residents and union workers bused in by the Mackles from Miami. Jammed into an auditorium at Lely High School near Naples, some wore hard hats and work boots and carried signs that read "Take the Shackles off the Mackles!" and "Permits Mean Jobs!" They booed and hissed, but Scott continued: "The right to invest in speculative ventures in a free market carries both the potential for profit and risk of loss." Therefore, "it is not reasonable or consistent with . . . American values to expect the government to assure anticipated profits by allowing activities which are clearly not in the public good."[85] Scott and representatives of other environmental groups and government agencies such as the U.S. Fish and Wildlife Service and the Florida Game and Freshwater Fish Commission denounced the project, which led local resident and Marco Island Democratic Club chairman Wilber Dickey to declare that they were "birdwatchers" and "termites."

"We want to see Marco completed as planned," said Dickey, but "do we respect law and order or do we change the rules midway and let a little group of termites come in and change things?"[86] Bob Georgine, president of the AFL-CIO's Building and Construction Department, claimed that upwards of 700 men "would have been working out there if the project hadn't been halted," while Ken Henning, an official with the Piledrivers Union local, blasted the Environmental Protection Agency (EPA) and the environmental movement by saying, somewhat artlessly, that "saving mangroves for children was bunk." Stoked by union leaders supplied by the Mackles and influenced, perhaps, by the *Naples Daily News*, a publication owned by the Collier family, the anger in the room was palpable. At one point, an unnamed union official demanded that each of the environmental groups present be investigated to ascertain if they were "communist oriented," while a shrill woman wearing a "Marco Island Now!" button yelled at an Audubon representative: "That's our island. Not yours!"[87]

It appeared for a time that the Audubon Society and a coalition of conservationist groups led by the Environmental Defense Fund (EDF) were willing to compromise. On day two of the hearings, they offered a plan by which Deltona would decrease the size of the development from 3,500 to 800 acres while building high-density, multifamily structures on elevated land away from the water instead of canal-front lots. Essentially townhouses, the structures in the EDF plan required the same number of workers as the Deltona plan but eliminated dredging and saved some 2,200 acres of mangroves and 735 acres of bay bottom. "In these meetings . . . we have made it clear," said Charles Callison, a vice president at Audubon, that "[we are] not against development. Housing needs must be supplied," but "there is a right place and a wrong place, a better way and a worse way, right here in Collier County." Nevertheless, both the Marco Island Civic Association and Deltona attorney Jack Peeples insisted that changes to the project were "not negotiable," for the simple reason that Deltona had presold some 5,325 lots. Peeples even told the crowd, to raucous applause, that under no circumstances would changes be "foisted" upon Marco.[88]

Tough words, but as the hearings ended in September 1975 and Deltona's dredge and fill application wound its way through the Corps, no fewer than nineteen environmental groups and six state and federal agencies opposed the permits, including the Department of the Interior, whose assistant secretary was now none other than Nathaniel P. Reed, and the EPA, whose chief administrator was now Russell Train. Both had tussled with Deltona before, but now Train, under the Federal Water Pollution Control Act Amendments of 1972, had veto authority over any dredge and fill permit submitted to the Corps that was deemed damaging to fishing, wildlife habitat, recreational areas, or water supplies. Thus, when Train threatened to veto the permits in March 1976, the Deltona application—in spite of high-level support within the Corps—was through.[89] In April, Lieutenant General W. C. Gribble Jr., chief of engineers, denied the Barfield Bay and Big Key permits outright, but chose to ignore the EPA in approving the Collier Bay permit, the rationale being that Deltona had already done work there, in upland sections, and that at least those buyers should get their lots.

Considering the magnitude of the permit denials, it was the least that the Corps could do. However, that still left some 4,307 refunds, a full $55.8 million in payments made on contracts-for-deed. "Almost every executive [we had] began negotiating with people to get them to settle on other property," said Frank Mackle III, because "our assets were tied up in land" and "couldn't be liquidated quickly. . . . It dealt us a massive blow." Citing a "ripple effect" in a brutal economy, in the coming years the company sold off some three miles of Marco beachfront, the Marco Island Hotel, and properties throughout Florida to settle some $200 million in debts. It also cut nearly 700 sales and management positions, shook up its foreign offices, and shifted away from installment land sales to finished homes. The economy had a lot to do with it, but the trouble began at Marco. "In hindsight, we shouldn't have sold the lands without the permits," said Frank III, ruefully. "But the state did not require you to have permits in hand before selling land, because permit approval was considered routine." That may have been true, but Audubon official Charles Lee put it another way: "They elected to shoot craps and lost."[90]

CHAPTER FIVE

# The First-Mortgage Frauds

> We started the $10 down, $10 a month deal, but it drew in a lot of bad people who abused the business.
>
> —FRANK MACKLE JR., President, Deltona Corporation

In late 1974, business writer Richard Dennis was hard at work in the newsroom of the Fort Myers *News-Press* when in walked an ink-stained, fifty-nine-year-old coworker named Phil Profita. A transplant from suburban Chicago, Profita worked in the pressroom, printing newspapers, and told Dennis that he was having trouble with a local brokerage firm called Homestate Investment Inc., so he asked Dennis, a former banker-turned-journalist, for help.[1] In 1969, Phil and his wife Anna had moved to Cape Coral from River Grove, Illinois, after Anna developed health problems that required a warmer climate. Phil, who'd been a machinist at the *Chicago Sun-Times*, found similar work at the *News-Press*, but money was tight and the couple feared that their nest egg wasn't enough for retirement. However, in 1972, Phil saw an ad in the newspaper that read: "This Offering to Residents of Florida Only. Corporate Notes Secured by First Mortgages Paying 12% Annual Interest. Monthly Interest Checks."

Below those words was a chart. Over forty-eight months, it read, a $5,000 investment paid $50 a month; a $10,000 investment paid $100 a month, and so on. There were "no brokers fees" and "no sales charges," and Homestate, the company selling the notes, was a "licensed and bonded" brokerage with offices in Lehigh Acres and Cape Coral.[2] Phil couldn't believe it. Twelve percent? His certificate of deposits paid maybe 6 percent, if that. Because Homestate looked like a reputable broker, Phil called the company to check it out. He spoke to a man named Jay Aten Jr., who explained that Homestate took a small commission for selling attractive promissory notes for a first-rate Miami developer named Continental Land Management, which owned two installment land sales communities near Inverness called Inverness Village and Citrus Park. Located seventy-five miles north of Tampa in a wooded region known for its fishing, Inverness was booming, he said, while homesites there, in just these two communities, had been appraised at $6,000 a lot.

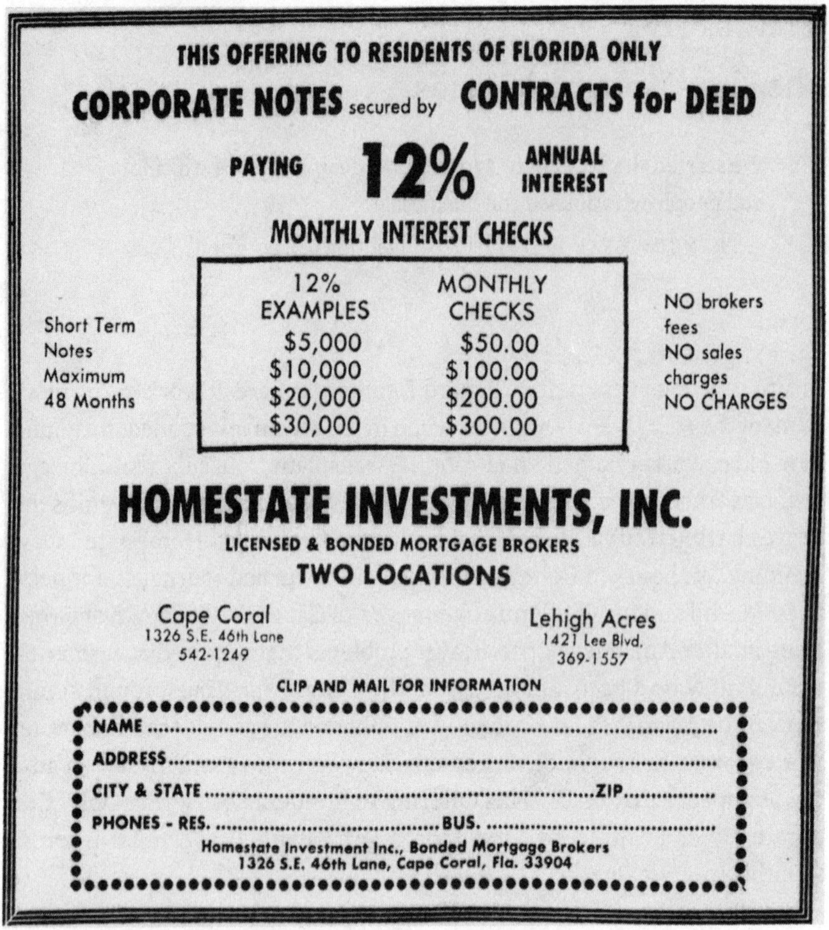


In 1974–75, reporter Richard Dennis uncovered a massive, $350 million securities fraud scheme in which sketchy brokerage firms promised retirees a 12 percent annual return on investments backed by installment contracts on lots in various Florida developments that the firms didn't own. Courtesy of the author.

To capitalize on this, said Aten, Continental came up with a plan. It would offer short-term promissory notes to ordinary men and women at an extremely attractive 12 percent rate of interest. In exchange, these ordinary investors would contribute at least $10,000, and after four years of monthly interest checks they would get their principal back.[3] In other words, $10,000 went out and over a four-year period some $14,800 came in. It was an incredible deal, but the beauty of it, insisted Aten, wasn't just the high rate of return, which doubled what banks were offering. It was that the notes were collateralized by valuable installment lots that Continental itself owned, but on which buyers were making payments, hence the term "first mortgage." In fact, Aten told Phil that Continental had a policy in which it backed every dollar invested with at least twice that in land, so if he gave the company $10,000, Continental would secure the investment with $20,000 in lots. Bottom line: the Profitas couldn't lose.

Thus convinced, in 1972 the couple handed Aten their life's savings, $21,000, and for more than a year, checks arrived like clockwork. But, in early 1974, with no warning whatsoever, they stopped, and in April, the federal Securities and Exchange Commission (SEC) accused Homestate, a different brokerage firm in Fort Lauderdale named Hartwell & Associates, Continental Land Management, and four other Continental-owned companies of fraud. A court named a receiver, and in the fall of 1974, Homestate and Hartwell lost their licenses while Continental went bankrupt.[4] That's when Phil Profita visited Dennis in the newsroom. The journalist listened to the story and then followed up, interviewing Profita's friends, who had also invested with Homestate. That led to more friends, more Continental subsidiaries, more land developments, and a complex web of loosely affiliated land companies and brokerages that had perpetrated one of the largest land frauds in American history.

In retrospect, Florida's "first mortgage" frauds of the 1970s were round two of a larger series of frauds that *Wall Street Journal* reporter Jonathan Kwitny described in his 1973 book *The Fountain Pen Conspiracy*, in which a loose ring of sophisticated hustlers, based (at one point or another) in Fort Lauderdale but with offshore links to a series of banks in Central and South America, the Caribbean, and Europe's Isle of Sark, perpetrated any number of scams, each more audacious than the last.[5] In 1968, for example, they formed a fictitious bank, the "Bank of Sark" near Guernsey, which used fraudulent accounting figures and professionally printed bank-to-bank cashier's drafts and fake mutual funds to steal some $40 million worldwide. One group set up a fake insurance company selling fake policies; another set up a money-order service that didn't send money; and still another made up a real estate agency that sold thousands of acres of property that people already owned. They sold nonexistent securities and mining claims and were adept at the "advance fee" racket, in which desperate businessmen, unable to secure bank loans in a tight lending market, turned to so-called money brokers for help.

For a fee, sometimes 8 percent or more upfront as a commission, the brokers would find companies with cash that were willing to lend the money. Such brokers were often legitimate, but in one example from 1969, the ring established a shady brokerage firm named Alabama Equity (AE) to secure an 8 percent fee, or $48,000, from a Seattle nursing home company needing $600,000 to expand. AE accepted the money, then connected the company to a supposedly legitimate operator named Cumberland Insurance Investment Group of Delaware, which had over $53 million in assets. But, what AE didn't say was that the same person who established the AE brokerage firm had established Cumberland, or that Cumberland had no intention of paying anyone anything, or that most of its $53 million had been guaranteed by the Bank of Sark. The AE scam netted $700,000 that Birmingham officials knew of, but there were other scams, in other states, that were incredibly difficult to detect, let alone prosecute, because of weak laws and an impenetrable web of brokerage firms, law firms, offshore banks, and subsidiaries. And this was just one ring.

In fact, in 1971, wrote Kwitny, a convicted securities vendor who sold promissory notes for a fraudulent religious foundation told the House Select Committee on Crime that upwards of 500 people were "hanging bad paper full time."[6] They congregated in Florida because of the state's proximity to offshore banks in the Caribbean and because Florida's chief regulatory agency, the state's Division of Securities, was notoriously lax. In 1976, for example, state auditor Ernest Ellison completed a three-year review of the agency, finding, among other things, that it had no written policies regarding background checks for individuals wishing to sell securities in Florida, nor had it spelled out its own investigative procedures. What is more, the securities division had just nine investigators total, who, overstretched and underpaid, policed tens of thousands of brokers who in 1972 did $12 billion in business.[7] Officials knew they were understaffed, but nevertheless, in 1973 state comptroller Fred Dickinson claimed that Florida's securities industry had "a reputation for honesty and integrity" and had made a "lasting contribution" to the state's "stable business climate."[8]

Never mind that former Fort Myers resident and "Fountain Pen" conspirator Paul Noe had just been convicted in Dallas for his part in a $273 million international securities scheme, which included Florida, or that schemes like his were surfacing all the time.[9] In 1973 and 1974, for example, authorities discovered a nine-person ring operating out of Orlando that used counterfeit securities to get bank loans; a firm in Fort Lauderdale that sold worthless municipal bonds to returned prisoners of war; an oil company in Tallahassee that peddled shares of nonexistent wells; and an evangelical minister in Dade City who floated 8.5 percent "church bonds" before running off with the money.[10] First-mortgage frauds became popular among swindlers after authorities wised up to the various "Fountain Pen" schemes: the creation of banks with no assets, the substitution of good stocks with bad or fraudulent stocks in company portfolios, and so forth. But the economy helped, too; it plummeted, with the S&P 500 stock index losing 45 percent of its value between early 1973 and late 1974, when inflation hit 11 percent.

Beset by what one congressional study called "runaway living costs," in 1973, a middle-income family with a household budget of $12,614, for example, had to pay an extra $1,168 to live like it did in 1972. Among other things, the family paid $502 more for food, $154 more for housing, and $57 more for clothing and personal care. Said one senator: "Most Americans found themselves on a treadmill in 1973. They worked more hours for more money which bought less for their families," and sometimes poor retirees had it the worst.[11] In Florida, where one out of five residents were age 60 and over, the average Social Security payment for a retired single worker was $162 a month and for a married couple $271. And, though many retirees had pensions, one estimate suggested that monthly Social Security checks made up 90 percent of the income for one out of every three.[12] "[A decade ago], Social Security was comfortable," explained Len Esch, a Florida Family Services official in 1974, "but this is no longer the rule. Everything has spiraled in cost," and "what was comfortable at that time is not nearly adequate now."[13]

Experts feared that Florida had so many retirees, an estimated 1.5 million by 1970, that planned communities such as Port Charlotte were becoming "retirement ghettos." There, some 57.3 percent of families received Social Security while just 40.3 percent received wages.[14] And, though some retirees worked, appearing in both categories, double-digit inflation made it harder to make ends meet, particularly when a person got sick or a spouse died. "Barryl and I looked forward to retiring and coming to Port Charlotte," said a Poughkeepsie transplant in 1974, so "[we] came south as quickly as Barryl retired. He died four months later. What money we had is tied up in the house and I'm having a hard time living on Social Security and a small pension. My children and their families are still in Poughkeepsie, and I can't even afford to go back home. . . . It's wonderful to be in the sun instead of the snow," but "we just don't have enough money [to live here]."[15] The headlines didn't lie: "Dreamers Find Paradise Costs More than Social Security Pays" (*Fort Lauderdale News*); "Elderly Poverty Level May Hit 45 Per Cent" (*Tampa Tribune*); "Wintertime—and the Living's Not Easy for Retirees on Social Security" (*Orlando Sentinel*).[16]

Thus, to make ends meet, Florida's elderly sometimes sold their houses, finding apartments or even cheaper accommodations in rented mobile homes. They took on roommates, applied for food stamps, crowded restaurants with senior citizens' specials, and—if they were physically able and could find them—worked second jobs. Retirees "do everything they can" to get by, said a social worker in Miami in 1971. "They buy three pounds of hamburger, and stretch it out all week, cooking it in different ways. They watch for sales on pot pies—one makes a meal for them with a little rice," and "they buy stale bread and overripe vegetables and fruits." Some even "sleep late to miss breakfast," she added, then "eat a big lunch" and "a little dinner" before going to bed.[17] Those who had nest eggs were lucky, but increases in living expenses and surprise medical costs whittled their savings accounts down to zero. For others, the money they did have was worth less, for $1,000 deposited in 1963 at roughly 4.7 percent returned $1,587 in 1973. But, with inflation, that $1,587 had the spending power of just $1,143.[18]

Stuck with such low yields, retirees began transferring their money in droves, from "passbook accounts" and certificates of deposit at local banks to higher-earning investments such as bonds, sold through brokerage firms. "I took [my] money out of the bank, where it was earning 6.5 percent," said a gipped first-mortgage investor from Merritt Island, Florida, "because [the investment program] was supposed to give me $100 a month on top of my Social Security. . . . That was my life savings." Of course, retirees had several different bond types to choose from, including AAA-rated corporate bonds or tax-free municipal bonds paying 8 or maybe 9 percent. But, with ads in Florida newspapers touting first-mortgage bonds at a whopping 12 to 14 percent, thousands of investors found securities backed by land deeds impossible to resist. Estimates vary, but in 1975, state comptroller Gerald Lewis stated that "conservatively," Floridians had invested $350 million in first-mortgage securities sold through fifty-six different companies whose holdings were in remote places such as Inverness.[19]

The seat of sparsely populated Citrus County, one of eight so-called Nature Coast counties north of Tampa Bay and south of the Panhandle, Inverness was known for natural springs, moss-covered live oak, lakes with bass fishing, Pentecostal churches, and the Ku Klux Klan. The area had been home to a booming citrus industry, then a booming phosphate industry, but by World War I growth had collapsed in Citrus and the population began to decline, from 6,731 in 1910 to 6,111 in 1960. Most residents were farmers or ranchers, while some impoverished families subsisted into the 1950s by gathering moss in bulk for use as stuffing in the furniture industry. Socially and culturally, Citrus belonged to the Deep South, so race was a touchy issue here. The county saw Klan rallies in 1957 and 1965, but managed to avoid much of the more grotesque violence associated with neighboring counties Levy and Hernando.[20]

When, in the 1960s, land giants like Deltona moved north into central and northern Florida in search of large tracts to develop, Citrus County became a prime locale. In 1961, the Rolling Oaks Corp. purchased 5,000 acres between Crystal River and Hernando for its "Beverly Hills" development, the so-called Alps of Florida, and in 1966 Deltona tried but failed to secure the 13,000-acre Black Diamond Ranch to the west of Beverly Hills from former state senator John Stansel Taylor Jr. The company tried again in 1969, when it successfully purchased an 11,000-acre spread from the Camp Phosphate Co. and several adjacent properties for its "Citrus Springs" community near Dunnellon, while in 1972, Punta Gorda Isles Inc. (PGI) bought some 14,700 acres near Homosassa for a sprawling development called "Sugarmill Woods."[21] By 1973, Deltona, Rolling Oaks, and PGI were the "big three" in Citrus, having sold tens of thousands of lots, but as retirees began favoring the area, other less reputable companies got into the act.

One of them was Continental Land Management Corporation, which seems to have been a bogus operation from the get-go. Founded in 1971 in Miami by a dubious certified public accountant named E. John Wentland, who in the 1960s had been caught falsifying inventory figures in an SEC registration statement filed by a Florida window company, Continental had no experience whatsoever in developing land.[22] Nevertheless, Wentland convinced (or hoodwinked) a Miami real estate firm that owned acreage west of Inverness to sell him the land on credit through a $772,000 promissory note, which it did sometime in 1971.[23] Soon after, and for no legitimate reason, Wentland formed four different subsidiaries to his Continental Land Management Corporation, including the Continental Companies Corporation, the Continental Land Corporation, Continental Land Development One, Inc., and the Continental-Southeast Land Development Corporation, which immediately offered, through the brokerage firms Homestate and Hartwell & Associates, 12 percent "first-mortgage bonds" backed by lots to which he didn't hold deeds. In other words, Wentland had sold second-mortgage bonds, not first, so the securities he peddled were worthless.

Of course, Homestate and Hartwell knew the bonds were fishy. Any broker would have, for a $10,000 investment required a $4,800 return in four years. But, since Continental paid commission rates of 20 percent to each firm upfront, that initial investment was in reality just $8,000. So, in effect, Continental had four years to turn $8,000 into $14,800, a seemingly impossible task requiring a jaw-dropping payout of 21.25 percent per year. And that's before Continental's own expenditures, and before it, too, earned a profit. Yet 20 percent was significant, so Homestate and Hartwell turned a blind eye to the business and sold $5.7 million in Continental securities in two years.[24] But Wentland wanted more. He used the sales as seed money to plat his two Citrus communities and to purchase land for an additional community called St. Johns Riverside Estates in rural Putnam County east of Gainesville. In all, he created 6,301 lots, then had two Houston, Texas, appraisers, who weren't even licensed in Florida, value the properties at an incredibly exaggerated, if not clearly fraudulent, $39 million, or some $6,200 apiece.[25]

He then ran ads, built model homes for visitors, and started work on a bare-bones condominium complex at Inverness. It seems counterintuitive that Wentland would have spent money to steal money, but first-mortgage scams like his required fledgling residential communities that at least looked legitimate, even if they weren't. Therefore, like any real developer, Wentland registered his property with Florida's Division of Land Sales, the successor agency of the Florida Land Sales Board, knowing full well that authorities there had no say over securities. Likewise, Wentland also knew that securities backed by first mortgages were exempt from oversight by the state's Division of Securities, while securities sold within Florida were exempt from the SEC, which is why Homestate's advertisements read: "This Offering to Residents of Florida Only." However, as Richard Dennis discovered, all three agencies were aware of the scheme as early as 1972, but first-mortgage bonds (if secured correctly) were legal, and none of the agencies had jurisdiction.

"The biggest reaction I got [from them]," Dennis told a Florida House committee in August 1975, was, "'it's not in our jurisdiction. It's in the jurisdiction of X.' But, on inquiring of 'X' I would hear: 'It's not in our jurisdiction. It's in the jurisdiction of Y.' So you go down the line. . . . But in a careful analysis and in coming to the law exactly as it should be interpreted, they were right. It was not their jurisdiction. And the schemes . . . dovetailed between [them]."[26] Nevertheless, things came to a head in 1974, when Continental Land Management made a fatal mistake. In contravention of federal securities laws, it sold some $240,000 in unregistered securities across state lines, which made it and its subsidiaries, Wentland, three Maryland brokers, and Homestate and Hartwell & Associates the subject of a full-blown SEC investigation. It seems they had sold securities backed by Continental's Inverness developments to two Maryland women—including one whose husband had died in a hunting accident and who invested her entire $100,000 insurance payout for her children—but had hid the transactions from the SEC by listing the women as Florida residents.[27]

They weren't, so when the women complained to the SEC that Continental had missed its interest payments, the agency used the violation to shutter the company's offices, confiscate its records, and lower the regulatory boom. The question was, Where did the money go? In 1974, a court-appointed receiver found that in just three years of business, the company owed at least $7 million in notes to investors and had "a deficit capital position" of more than $3 million.[28] But no one knew what it owed. The records were incomplete. Wentland held management positions in shadowy first-mortgage companies outside of the Continental group, including one called the Stuart-Martin Development Corporation in Stuart, Florida, which itself bilked $4.3 million from investors then transferred hundreds of thousands of dollars to Wentland in the form of a loan.[29] Much of it disappeared, offshore, while Dennis and Fort Myers–based State Attorney Joseph D'Allesandro and Assistant State Attorney Louis St. Laurent linked several Wentland associates to organized crime.[30]

As for the investors, the SEC found a thousand or so Continental securities owners when it closed down the company in 1974, but since Continental didn't own the properties its securities were based on, they lost every dime. "It's all gone," Phil Profita told Dennis, and "most of our friends and acquaintances have been hurt, too."[31] That was the gist of the story when Profita visited Dennis in late 1974. But as Dennis investigated the Continental case and contacted officials at the Division of Land Sales, they told him they didn't know anything. They were "noncommittal," he said, and gave "the impression that they really were not aware of such programs to any significant degree being proliferated in the state."[32] However, in early 1975, someone sent a manila envelope to Dennis at the *News-Press*. It was unmarked and full of Division documents revealing that not only had the agency known of the scams, it knew of the scope but had been "either inexcusably negligent of investors' welfare or purposely helpful to companies that were bilking the investors" themselves.[33]

At issue was whether or not the Division of Land Sales could have stopped first-mortgage security sales by simply refusing to register companies that used it as a form of financing. The Division definitely could have, but fearing lawsuits and having a close relationship with developers and a company called Viking General in particular, it did nothing. Viking, a Miami firm that heretofore had specialized in printing sales literature for other land companies, had sprung up seemingly out of nowhere by acquiring the 50,000-acre

Griffith Ranch in uber-remote Okeechobee County in 1972. Located north of Lake Okeechobee at the eastern end of the "Ninety Mile Prairie," the Griffith Ranch encompassed mile upon barren mile of seasonally inundated cattle land, which, because it was zoned agricultural, was largely undevelopable. Nevertheless, Viking bought 6,000 acres for $1.5 million, then optioned the remaining 44,000 acres at $245 apiece. That was the beginning of "Viking Country," today's Viking Estates, where buyers paid $70 down and $70 a month for 1.25 acres of nothing.[34]

Even so, and being painfully aware of the problems it had with similar raw acreage developments like River Ranch, the Gulf American boondoggle, the Division approved Viking Country and even this fanciful script that forty-five salesmen on twenty WATS lines read with gusto every day: Viking Country "is ideally-located in the fastest-growing area of our state. There are so many things happening in this area I can't begin to tell you. And from a price standpoint, I feel this is the most outstanding purchase I've seen in Florida in the past ten years."[35] The script worked, because by 1973 the company had sold thousands of lots, then expanded quickly into mobile home parks and condominium communities in southern and central Florida as well as in Quebec. It borrowed millions, but when sales slowed with the OPEC oil embargo and the recession, it turned to securities, offering first-mortgage bonds through brokerage firms at 11 to 13.25 percent.

While certainly an upstart, Viking was a legitimate land giant. It may have skirted sales laws and fibbed to its customers from time to time, but unlike Continental and other first-mortgage securities dealers, its goal in raising money wasn't to steal it, per se, but to finance the development of new land. To that end, it played hardball with the Division of Land Sales by hiring as its legal representative Jack Graff, a friend and former law partner of then-governor Reubin Askew, who, critics alleged, either finessed or pressured the Division into letting the company's first-mortgage program proceed.[36] But, when the Division let Viking proceed, it let everyone proceed, including the Continental group and a company called the Equitable Development Corporation of Miami Beach, the developer of Paradise Acres near Mayo, Florida, which sold some $3 million in promissory notes to investors even though officials in Ohio had banned the company from doing business there because Equitable had gone bankrupt.[37]

Later, the state attorney general's office in Fort Myers linked Equitable to lawyer Albert George Siegel, "believed to be a nephew of Bugsy Siegel," as well as C. W. Deaton, a convicted conman who provided the company with bogus guarantees from a Cayman Islands surety firm that didn't exist.[38] (Deaton, by the way, had been a central figure in the "Fountain Pen Conspiracy," having sold fake silver mine securities in Texas and pilfering a bank he ran in Colorado before fleeing warrants in Florida, Germany, and Canada to live in the Caribbean.) Yet, for whatever reason, the Division of Land Sales proved incapable of connecting the dots. But Dennis could. He noticed, as did Assistant State Attorney Laurent, a Canadian transplant and grandson of former Canadian prime minister Louis Stephen St. Laurent, that many of the Florida companies that had been selling bad securities were also active in Arizona. They included Buckeye Mines, Inc., the 2609 Corporation, and three other companies with Arizona-related names: the Arizona-Florida Development Corporation, Arizona Florida Equities, and Corona de Tucson, Inc.

Digging further, Dennis learned that each of the companies had been headed or managed by a one-time Wentland associate named Robert E. Brown. In 1970, Brown, a Miami resident, acquired some forty lots in a struggling development in Volusia County, near Deltona, then used forged deeds and other spurious documents to sell 165 of them.[39] Somehow he evaded prosecution and used the money to purchase property in Arizona in a desert development called Lake Mead Rancheros. Founded in the early 1960s, the Rancheros were the brainchild of a Miami fraudster named Dory Auerbach, who duped northerners into thinking they were buying developed land. The Rancheros had no water, no roads, and no electricity, while Lake Mead, the supposed centerpiece of the development, was forty-five miles away. In 1968, Auerbach received a one-year suspended jail sentence and a measly $10,000 fine but retained 70 percent of the development and, in 1970, sold parts of it to Brown.[40]

That was the beginning of Brown's Arizona activities, which grew to include homesite and first-mortgage sales at the Rancheros and a second development in a remote area south of Tucson called Corona de Tucson. Corona, Dennis discovered, was a failed Baptist retirement community that Brown had purchased (on credit) in 1971 and whose lots he sold through a series of Arizona real estate firms, including one called Invesco Realty. Dennis linked Invesco to organized crime figure Ned Warren Sr., the "godfather of Arizona land fraud," which made him wonder what, if any, role Warren had played in Florida's first-mortgage frauds. It wasn't a smoking gun, but Invesco sold lots for Corona and Brown sold securities in Florida backed by Corona land. "Had Warren devised this thing?" Dennis asked. The *Wall Street Journal* said no, insisting that a broker named H. J. "Jim" McDonald of Rush, Minnesota, had done it first, in the 1960s, and had brought the scam to

Miami, where he taught it to an unethical accountant named David Edstrom, who later sold 2609, Continental, and Equitable securities and who in turn taught it to Brown.

But, by the early 1970s, Arizona officials had linked McDonald to a series of Phoenix companies as well, and behind those companies, Dennis hypothesized, maybe, just maybe, was Warren. In 1975, he flew to Arizona to check it out. Amazingly enough, Dennis had been a reporter for less than a year. His initial first-mortgage article had appeared in November 1974 and it was a short description of how a judge forced Continental into receivership. Then in January 1975, he wrote a multipage special with a headline that read: "Life Savings Lost, Probably Forever."[41] In it, he discussed the Profitas and other victims, the nuts and bolts of the scam, and what officials in Tallahassee were doing about it—which to that point was nothing. Picked up and syndicated nationally by the Gannett News Service, the reporting struck a nerve. The state's Division of Land Sales and Division of Securities were both inundated with complaints, while the state attorney's office in Fort Myers received 850 calls in a week.[42] Dennis kept writing, decrying what he called the "apathy and indifference" of state regulatory agencies, which, he insisted, were "ineffective, unworking," and "abusive to the public good."[43]

When, in January 1975, Dennis confronted Robert Shevin, the Florida attorney general, at a banquet in Fort Myers and then wrote about it, the state's chief lawman finally began to act, joining newly elected state comptroller Gerald Lewis in forming a "Land-Fraud Task Force" of district attorneys and representatives of state and federal regulatory agencies who vowed to share information to pursue wrongdoers across jurisdictional lines. "I'm serving public notice that we're going after these people from every angle and in every way that we can to put a stop to this [thing]," said Lewis, who promised to give "top priority" to prosecuting "companies and individuals" that had been selling fake securities.[44] That was in May 1975. But, by this point, news had leaked (to Dennis and others) that both the Division of Land Sales and the Division of Securities had been aware of the scam and done nothing, and that Shevin's own assistant had written a report on the subject in 1974—though Shevin denied ever reading it. Hearings ensued, several land sales officials resigned or were fired, while the majority leader in the Florida House of Representatives, Dick Clark, blamed the fiasco on Lewis's predecessor Fred Dickinson.[45]

As for the Land-Fraud Task Force, its most notable success involved naming, publicly, the fifty-six companies it alleged had perpetrated the scam and pulling the licenses of more than 200 unscrupulous brokers. It also provided information to and assisted state and federal prosecutors in convicting, among others, E. John Wentland, Lane Hartwell, David Edstrom, and Jay Aten Jr., as well as two father-son teams: Bernard and David Horowitz, and Wallace White Jr. and Sr., whose Continental, Equitable, Stuart-Martin, and 2609 Corporation scams had bilked investors of at least $20 million. Then, in August 1975, Florida Attorney General Shevin announced a joint investigation with Arizona Attorney General Bruce Babbitt, who explained: "We both have the same job in the two most volatile securities and land sales states in the nation, and we're learning that the people involved in each state are pretty well connected with each other."[46] Exhibit A, wrote Dennis, was Corona de Tucson developer and one-time 2609 Corporation president Robert E. Brown, as well as Ned Warren, the land-fraud superstar who had actually lived in Florida before moving to Arizona in 1961.

Born Nathan J. Waxman to a wealthy New York family in 1914, Warren left the University of Pennsylvania as a freshman to spend most of his adult life committing crimes. He started small, earning a "theft by false pretense" charge at age 22 in Providence for accepting $339 in donations for a nonexistent "Blind Musicians Group." Other arrests followed, including one for fraud in Chicago, where in 1946 he received two years' probation for attempting to pass $15,000 in bad checks. In 1949, in New York, the seemingly relentless Warren cheated investors out of $39,000 for a Broadway show that never happened, which landed him in Sing Sing prison, then in 1955 he relocated to Florida, where he scammed residents in a chinchilla breeding scheme and also sold raw investment acreage in the Everglades. But, before Warren did too much damage, at least in Florida, in 1957 a federal court sentenced him to four years in prison for concealing $25,000 in assets in a bankruptcy proceeding. Released in 1961, he moved to Arizona and reentered the land business.[47]

There, he formed one installment land sales company and mortgage and realty company after another, more than a hundred in all, which, beginning with the Thunderbird Land Corp. of Phoenix, sold 1½-acre tracts of barren scrub brush for $795, or $10 down and $10 a month.[48] Like the Rosens in Florida, Warren used call centers and cattle chutes and boiler rooms, and he and his salesmen played every trick in the book. But whereas the Rosens ran one large company, the Gulf American Corporation, Warren ran dozens, moving cash and contracts and property through a complex web of corporations and partnerships that even seasoned accountants had trouble figuring out. Warren seldom, if ever, held positions in companies on paper—usually a criminal named Tony Serra or Warren's cellmate at Sing Sing, Bill Steuer, did—explaining how, in part, Warren's Cochise College Park development could bilk investors of $38 million while Warren was never charged.

"Until recently, Warren operated with impunity," wrote a group of investigative journalists in 1977. "While a few associates went to prison, Warren darted from one corporate entity to another like a broken field runner, several steps ahead of prosecutors."[49] He evaded perjury and bribery charges in 1974 and 1975, respectively, when evidence went missing from a locked police cabinet and a witness named Edward Lazar, an accountant and former president of Consolidated Mortgage Corp., a Warren company, was shot four times in the chest and once in the back of the head, gangland style, in a Phoenix parking garage.[50] Lazar had been set to testify against Warren and to say that he had made payments to various Warren-controlled firms, which Warren then used as bribes for J. Fred Talley, the former head of the Arizona real estate department.[51] Nevertheless, the police failed to tie Warren to the crime, and no one was ever charged. But this was the seventh instance since 1971 in which someone close to the developer had died. There were two plane crashes, two heart attacks, a suicide, a shooting, and a car crash, each suspicious and each advantageous to Warren.[52]

However, in September 1975, a U.S. district court in Seattle convicted Warren along with son-in-law Gale Nace of extortion for threatening to break the arms and legs of a bar manager who owed the developer $8,000. Warren received two twelve-year concurrent sentences, but he posted bond during the appeals process and spent 1976 at home in Phoenix, hoping to entice state and federal officials with a plea deal.[53] He was sixty-one, had cancer, and had heard that the Maricopa County attorney's office planned to charge him with at least twenty counts of grand theft for misrepresenting land he sold to U.S. servicemen abroad. Facing decades in prison, on top of a twelve-year extortion sentence, Warren would name names. Or would he? According to the FBI, he had also acquired a forged passport and had moved what the FBI called "large sums of money" offshore. So was he ready to run? Or was he looking for a deal? And if so, from whom? The feds had cases open that he knew about, in Florida, Arizona, Utah, and New Mexico, and officials believed he had corrupted everyone from J. Fred Talley in the real estate department to county attorneys and state representatives and even a sitting judge.

Violent, crafty, and shrewd, Warren was a supremely intelligent career criminal who thought two or even three steps ahead and who, in May 1976, asked Richard Dennis to meet him. No one knows why he did, and Dennis, who died in the 1980s, never said. But Dennis had been on assignment at Gannett's Rochester, New York, headquarters and in late 1975 had published a syndicated article on how first-mortgage frauds had spread to New Mexico. He mentioned Warren in passing but wrote that "the same people and the same companies" that had bilked investors in Arizona and Florida were "either indirectly or directly involved."[54] Was Dennis writing a larger story that implicated Warren that Warren wanted to know about? Or was Warren hoping to use Dennis as a conduit to prosecutors in Florida in hopes of striking a deal? It could have been both of those things, but all Dennis said after meeting with Warren in Phoenix was that Warren had "asked for assistance that I could not provide."[55]

Less than a week later, in a seemingly unconnected development, veteran *Arizona Republic* reporter Don Bolles, who had first "outed" Warren as a con-man in a series of articles in 1967, was seriously injured and later perished when a bomb blew up in his car. Bolles, in one of his last utterances before dying, named as his murderer John Harvey Adamson, a low-level hooligan and one-time Warren associate who had tricked Bolles into meeting him by claiming he possessed evidence of a fraudulent land deal involving high-level Arizona Republicans, including Senator Barry Goldwater. Adamson didn't have any evidence; the meeting was a ruse, set up to draw Bolles to a Phoenix hotel while Adamson wired the bomb outside. It exploded shortly before noon on 2 June 1976, a Wednesday, as Richard Dennis was across the country in the *News-Press* newsroom. As Dennis later explained, he had no idea Bolles had died when the phone rang. The caller refused to give his name but said tersely: "We flipped a coin, and you won."[56]

Only later, when Dennis heard of the Bolles's bombing, did he understand what it meant. Warren was threatening him. Or someone else was, someone in or with ties to Arizona who knew of the bombing before even Dennis did and who wanted the reporter to stop. The odd thing is, when Adamson pled guilty in 1977, prosecutors pinned the bombing not on Warren but on a wealthy lawyer named Max Dunlap, who, they alleged, had paid Adamson to kill Bolles for a series of articles he had written on Dunlap's friend and business partner Kemper Marley Sr., a cattle and liquor magnate who supplied alcohol to the state's dog-racing tracks while sitting, in a conflict of interest, on the tracks' politically appointed oversight board. Marley resigned, but Dunlap, prosecutors said, paid Adamson to kill Bolles in retaliation. But Dennis didn't know that, not in 1976. Therefore, he took the call as a warning. Bolles had seven children, as Dennis himself did, so a week later the upstart

Florida reporter, the recipient of three Pulitzer Prize nominations in two years, resigned.[57] He never worked as a journalist again.

RICHARD DENNIS LEFT the *News-Press* in June 1976, less than a week after the Florida legislature passed a series of land fraud protection bills that effectively ended first-mortgage securities frauds. By law, all mortgage securities now had to be registered with the Division of Land Sales while second mortgages promoted as first mortgages were illegal. In addition, the legislature strengthened the Division's oversight over all installment sales by requiring developers to disclose the appraised price of the lots they were selling, which had never been done before; to set aside trust accounts for customers who'd changed their minds within thirty days or who'd visited the site and changed their minds within six months; and to use a uniform accounting method so that regulators knew they were solvent. "Henceforth and forever more shall Florida have its blackeye removed for protecting the developers and letting the consumer get hooked," proclaimed Representative Paul Nuckols of Fort Myers. "These rackets have been going on in Florida since the 1920s. Now the people of the nation can have confidence when they buy [Florida] land."[58]

However, Florida's first-mortgage frauds hadn't killed people's confidence in buying land; neither had hearings (both in Washington and Florida), lawsuits, convictions, "buyer beware" statements by, for example, the Better Business Bureau, or nationally syndicated news reports by journalists such as Trevor Armbrister, David Langford, Robert Caro, Richard Dennis, Elizabeth Whitney, Morten Paulson, and Anthony Wolff, among others. In 1972, the *Stars and Stripes* European edition issued a three-part, twenty-four-page exposé on the industry's sales-to-soldiers business. Paulson and Wolff both wrote books on the subject: Paulson's *The Great Land Hustle* appeared in 1972, and Wolff's *Unreal Estate: The Lowdown on Land Hustling* in 1973. Even *Sports Illustrated* warned buyers when, in 1973, its longtime fish and game reporter Robert H. Boyle wrote "Buy Now and Cry Later," a vicious takedown of the installment land sales industry in which he interviewed both Vince Conboy and Elizabeth Whitney and wrote "Florida is crisscrossed with paper lots that are either under water or unreachable or hold no likely prospect of development," yet, "the state has spent more money to protect alligators than it has to prevent buyers from being devoured by salesmen."[59]

Nevertheless, and in spite of the warnings, buyers made some $2 billion in installment payments in Florida in fifteen years, stopping only when the recession hit and the business itself became untenable. "What once was a flood has now become a trickle," wrote *Miami News* business editor Larry Birger in 1977. Some twenty years ago, "when Miami's Mackle Brothers began selling lots at Port Charlotte" and staged "their first national ad campaign in *Life* magazine, they were swamped with inquiries." Soon, "Florida land sales on the installment plan became a national industry." But, "after fifteen years of uninterrupted growth, land sales began to slow in the early 1970s, and five years later are at their lowest ebb since the industry was launched."[60] Like Deltona, many land sales companies diversified, or at least tried to; some others, such as the Royal Palm Beach Colony and the American International Land Corporation, decided to leave the retail business altogether; while in 1975, the giant GAC Properties, the Gulf American successor and owner of developments at Cape Coral, Golden Gate, River Ranch, Remuda Ranch, and Poinciana, as well as a mobile home community at Barefoot Bay, went bankrupt.

So did Cavanagh Communities, the owner of Rotonda West in northern Charlotte County near El Jobean, which the Deltona Corporation agreed to manage, but did not buy, in 1976. Opened in 1970 by a diminutive former salesman at the Gulf American Corporation named Joe Klein, Rotonda was a circular planned community on 5,400 acres of scrub brush that Klein's Cavanagh Communities purchased from brothers Alfred and William Vanderbilt, descendants of Commodore Cornelius Vanderbilt, in 1969. Klein, who cut his teeth in the business by selling television service contracts in Pittsburgh, had spent just five weeks with the Rosens before starting his own firm, the Worldwide Realty Company of Miami. The company specialized in selling other people's property—it wasn't, at first, a developer—and Klein would insist that some months, it sold more Gulf American property than Gulf American did. That was probably a stretch, but like Leonard Rosen, Klein was a salesman-savant who could, the saying went, sell water to a drowning man.

Even Elizabeth Whitney was impressed, albeit grudgingly, when in 1972 she met Klein (and his lawyer) for a report she was writing on Rotonda. He's "a polished and articulate man," she wrote, one "who flips the perfect word into precise usage as deftly as a postal sorter tosses letters into their proper cubbyhole. He smiles when it's appropriate and his mind is nimble and ingenious."[61] It was so ingenious, in fact, that in 1966 Klein somehow convinced John D. MacArthur, the billionaire insurance magnate and Florida property tycoon, to provide him with $70 million in land inventory in rural Martin and Palm Beach counties, as well as start-up funds for a new firm, Palm Beach Investment Properties (PBIP), to sell it. The land that PBIP sold was swampy and inaccessible; the 22,500 acres had no roads and was made up of an unimproved, raw acreage subdivision called Palm Beach Heights. And, at the center of Palm Beach Heights was Klein's baby: a planned city of some 60,000 people called Rotonda, the "City-in-the-Round."

Rotonda was nothing more than a sales pitch. It had been designed by a Klein associate named Charles Prynne Martyn, the founder of the Tequesta golf course community in Palm Beach County, who was neither an architect nor an engineer and who seems to have had no idea what his 5,400-acre planned community entailed. His concept was this: a 100-acre central core with a park and a lake with a tower in the middle. Then, radiating outward, a circular business district with 600,000 square feet of department stores, grocery stores, banks, restaurants, and so on, followed by eight 640-acre, wedge-shaped residential districts, each with a 7,000-yard PGA golf course and a marina. The coup de grâce was a 100-foot-wide outer ring and canal (fed by numerous other canals) that would drain Rotonda, as well as all of Palm Beach Heights, into the Okeechobee Waterway, meaning Rotonda residents could take boats from behind their houses and go on fishing trips to both the Atlantic and the Gulf.[62]

But, while Martyn claimed that "spiritual guidance alone" had "conceived [his] creation," he failed to consider the environmental backlash that draining the development would cause; so had Klein, who, reminiscent of the Mackles, began selling lots before the permits were approved. Between 1966 and 1970, PBIP and its retail subsidiary Palm Beach Heights Development and Sales Corporation would sell 8,127 lots for upwards of $48 million. They used "bird dogs," cold calls, plane trips, and cattle chutes—anything the Rosens did, Klein and his salesmen did, too. Author Jack Alexander, in his 1995 work, *Rotonda: The Vision and the Reality*, tells the story of George Perry, a Michigan football coach who visited Florida for a game at the Orange Bowl. Perry stayed at the Fontainebleau in Miami Beach and told Alexander: "The phone rang early the next morning. It was the hotel's social director asking if we'd like to tour a warship in Fort Lauderdale, then go inland and see some alligator wrestling."

Perry accepted—the tour was free, after all, and concluded with a cocktail reception at a hotel in nearby Hollywood, Florida, which featured a sales pitch. He bought "several lots," he said, but after returning to Michigan was surprised to discover a salesman at his home trying to sell him even more lots.[63] This was classic Klein: Both he and his sales force were extremely aggressive and frequently told customers that Rotonda was a done deal, even when it wasn't. In fact, as hearings to drain the property stretched on and on and grew increasingly acrimonious, especially when conservationist groups got involved, Klein continued to wow customers with a twenty-one-foot-diameter, room-sized replica of Rotonda that cost $40,000 and took a team of workers four months to build. It "had thousands of tiny houses," wrote Whitney, "cars in the streets, boats in the canals, trees on the golf courses. Intricate electrical wiring was installed . . . [to] points of interest so they flickered on and off like Christmas tree lights. Illuminated, the Rotonda model [made] New York at night from the air look like a brownout."[64]

When, after salesmen used the model to convince a wealthy Norwegian customer to spend a jaw-dropping $392,000 and the customer realized his mistake and complained—charging "misrepresentation"—the Florida Land Sales Board "ordered the room with the model locked."[65] It didn't matter. Salesmen showed it to customers anyway and would stand at the door and say, "I shouldn't be letting you see this, but. . . ." Those "little peeks," a salesman told Whitney in 1972, "clinched a lot of sales." They clinched so many sales, in fact, that PBIP, a company founded in 1966, became one of the hottest land companies in the state by 1969. That's when the Cavanagh Leasing Corporation, a publicly traded lumber, recreational home, and industrial equipment conglomerate, acquired PBIP for some $60 million in stock.[66] It was a partnership more than a purchase, really, as Klein emerged as president of a PBIP successor company called the Cavanagh Communities Corporation, which, under Klein's leadership, swiftly become a land giant.

In quick succession, it acquired two land sales firms in Ohio, a development company and a construction company in California, a modular home manufacturer in Indiana, and a marketing company in Florida called Miami Beach Vacations, a firm founded by first-mortgage scammer Bernard Horowitz that specialized in "free" sales trips. But first, Klein had a Rotonda problem. The Martin County Commission, the Florida State Board of Conservation, and the Central and Southern Florida Flood Control District all vehemently opposed the project, as did twenty-two different conservationist groups, the nearby city of Palm Beach Gardens—which, incidentally, was a John D. MacArthur development—and even the aerospace firm Pratt & Whitney, whose jet-engine testing facility to the south of Rotonda produced a deafening noise that experts claimed would render the development "uninhabitable."[67]

The Corps of Engineers held hearings, but before it issued a decision, Klein moved his Rotonda development across the state to Charlotte County, to the Vanderbilt's "2-V" cattle ranch on the Cape Haze Peninsula near El Jobean, which he renamed "Rotonda West." Some 25,000 acres in size, the land at Rotonda West was denuded scrub brush, used for timber and turpentine, which brothers William and Alfred Vanderbilt purchased in 1951. William had been the governor of Rhode Island; Alfred the owner and breeder of several famed racehorses, including Native Dancer, the Preakness and Belmont Stakes winner. Both were sportsmen and came to Florida on fishing trips to the posh and blue-blooded Boca Grande Hotel on nearby Gasparilla Island, which is where William became interested in a ranch. Therefore, he scouted the area with wife Anne, finding that acreage on the Cape Haze Peninsula, which had few roads and just a smattering of people, was incredibly cheap, so he and William, as partners, purchased a 35,000-acre parcel from a Florida real estate corporation for $700,000, or $20 an acre.[68]

With help from the locals, the Vanderbilts quickly established the 2-V Ranch, planting Bahia and Pangola grass and buying 3,000 head of range cattle purchased in part from A. C. Frizzell, whose ranch east of the Vanderbilts' eventually became Port Charlotte. They also grew watermelons and drilled for oil and in 1952 dredged over a million cubic yards of fill dirt to create a 275-acre subdivision called Cape Haze.[69] In the coming years, William would live at Cape Haze while running the ranch and building a successful business. However, by the late-1950s, the corporate push to subdivide and sell Florida meant that previously cheap ranchland had exploded in price, so much so that in late 1958, the Mackles offered the Vanderbilts $4 million for a 10,000-acre piece of the property bordering Port Charlotte. The Vanderbilts accepted, then ten years later sold the rest of the ranch to Klein in a bidding war between Cavanagh Communities and the parent company of Lehigh Acres for $19.5 million.[70]

Klein had only seen the property from the air, but with Martin County officials refusing to budge on what was now called "Rotonda East," he quickly needed an alternative site. As it stood, the Rotonda East property hadn't been platted. It was a come-on like Golden Gate, the planned community the Rosens had used to sell unimproved acreage in Golden Gate Estates. Though Klein had wanted to build his "City-in-the-Round," when it tanked, the 8,127 property owners who had purchased raw Palm Beach Heights acreage were out of luck. They'd expected a city, and most knew nothing of drainage or land-use restrictions when they bought the property, while some thought they could build there. Plus, every single customer had overpaid. An Ohio woman named Margaret Burroughs spent $7,000 on her property but received a tax valuation of just $1,700. "The land is in the Rotonda area," explained Martin County tax collector E. F. Schnursbusch, "and there is no way to get to the property except by boat. I wrote her that the assessment had been set at what we assume is the fair-market value until development creates a higher value."[71]

But that was the point: There'd be no development, ever, so Klein picked up his stakes and left. If buyers were misinformed or wanted roads to their property or a canal network, that was between them and the county. Because, as Klein insisted, he had nothing to do with it. "No doubt people expected eventually to have some value for their purchase," he said, "but there was no imputation of value or reference to future use by us. It was to be held for future use as determined by the county."[72] In other words, Rotonda East wasn't his headache, so in September 1970, Klein, undeterred and unblemished, opened Rotonda West. Emceed by *The Tonight Show* star Ed McMahon, whom Klein described as a "substantial investor" and Cavanagh's "vice president of community relations," the opening drew 12,000 people. They ate hot dogs, listened to speeches, and heard plans for a development that looked nearly identical to Rotonda East: a 5,400-acre, circular community with a park and a tower and eight emanating neighborhoods, except it would have seven golf courses instead of eight.

Land for the eighth course was a spring with a creek, which at least on one occasion Klein falsely claimed would provide boaters with an outlet to the Gulf.[73] It wouldn't, because to connect the two, Klein needed a deepwater canal that, if built, would contaminate Rotonda West's thirty-two-mile-long freshwater canal network, as well as its lakes and wetlands, with saltwater from the Gulf. Neither the state nor the Corps of Engineers would permit this, but as far as Klein and his salesmen were concerned, access to the Gulf was always implied. In 1972, for example, Cavanagh printed a full-page ad with a painting of a bird's-eye view of Rotonda in various Florida newspapers, such as the *St. Petersburg Times*, the Fort Myers *News-Press*, and the *Miami News*. The development stood at the bottom looking west and included at the top a wedge-shaped basin with a wide canal emptying into the Gulf. You couldn't miss it, but in tiny print at the bottom of the painting were the words: "True architect's rendering—not all improvements shown presently exist."[74]

Rotonda West was supposed to be "a dream community," wrote Jack Alexander, but "many early buyers heard only the appealing verbal sales pitch, saw only brochures the color of liquid sunshine selling them a vision beyond reality. They didn't read the fine print."[75] Even so, the overly ambitious Klein promised buyers a completed wheel of finished and improved homesites with electric lines, water lines, a sewer system, and drainage, plus seven marinas, seven golf courses with club houses, and a "navigable" canal network by the end of 1977. He couldn't possibly deliver, particularly when his focus wasn't on infrastructure building but on lot sales. With offices in forty-six cities, and one even in Switzerland, Klein and his salesmen were lot-selling machines. In fact, Klein was so eager to sell that in October 1969, on the afternoon officials in Charlotte County approved Rotonda's plats, he had his company secretary fly to Tampa in a jet to register the plats with the Florida Land Sales Board.[76]

Klein then hired former Gulf American associates as his executives, expanded his tourist-solicitation list to include bellhops, hotel managers, tour guides, desk clerks, and even pool attendants at no fewer than 100 Florida locales, and in February 1973 teamed with ABC Sports producer Roone Arledge to host the nationally televised "Superstars Competition," a multi-event mini-Olympics in which elite athletes from ten sports vied for $122,000 in prize money and were given free Rotonda lots. Featuring Johnny Bench (baseball), Jean-Claude Killy (skiing), Joe Frazier (boxing), and Rod Laver (tennis), among others, the events were held at what was a barren development of roads and some homesites but at most just 400 homes. Nevertheless, Klein jumped at the opportunity, for in exchange for hosting the event and housing and feeding athletes, Rotonda West would serve as the backdrop for one of the most popular shows on television, the *Wide World of Sports*.

For that, however, Klein and his construction crews, including hundreds of specially hired subcontractors, worked day and night just to finish the condominium complex where the athletes were staying. There were no hotels, so everyone else, including even Cavanagh officials, stayed in either Port Charlotte or Sarasota. "It's amazing," said ABC announcer Dick Button. "Last week we went to dinner in the restaurant here, and when we came outside the ground had been landscaped."[77] The development had just one finished tennis court for the tennis competition, a two-lane portable bowling alley for the bowling competition, and so forth, but Klein and his salesmen expected 15,000 people to attend. They didn't. Rotonda was just too remote, the events were held on weekdays when people were working, and with temperatures in the forties because of a rainstorm, it was freezing.

Tom McEwen, the folksy former sports editor at the *Tampa Tribune*, was there to see it. "At the appointed time for the first annual Superstars mini-Olympics," he wrote, "the weather was absolutely lousy. Yet, a knot of 300 perhaps gathered at 9:35 A.M. near the only finished tennis court" that "a helicopter had hovered over to dry. . . . Then came a prayer, the flag raising, and Ed McMahon stepped forward to read Governor Reubin Askew's proclamation [that 18–25 February was, in the State of Florida, 'Superstars Week']. 'Just wind me up, point me in the right direction and I'll nod and say hello,' said McMahon. It was cold and understandably tough to be funny. He tried a couple of more times with a couple of [beauty] queens . . . holding a ribbon somebody cut, and the first event was on."[78]

A celebrity Olympics with events such as golf and bowling seemed like a good idea, but the participants were bored, some said lackadaisical, while fans chose not to attend. In fact, Cavanagh had to bus the audience from event to event and put its own workers in the stands just to televise a crowd. Johnny Bench drank whiskey on the par nine. Jim Stefanich, a pro bowler, whacked a tennis ball into a canal, while Joe Frazier nearly drowned himself in the 100-meter swim meet. And what with the rain, Klein and other Cavanagh execs feared a public relations disaster. Indeed, they expected it. "Cavanagh worked so hard on this," lamented one official, and "we thought of everything. [We even had one guy] in charge of athletic supporter distribution. . . . We've been that thorough. Then we get this lousy weather. I'll bet Frank Mackle gets his jollies when he sees us on TV."[79]

But, a curious thing happened. When ABC broadcast the event the following weekend, splicing two windswept days into a show lasting two hours, the Superstars were a hit. The program won its timeslot and garnered an audience of 9.4 million homes, the equivalent of a Sunday NHL game and a Sunday NBA game combined.[80] There were, in addition, hundreds of datelines that read "Rotonda West" in newspapers throughout the country, whose journalists described the event as quirky, engaging, and—good for Rotonda—worthy of a watch. Therefore, Klein wisely agreed to sponsor the event again, this time with first and second rounds, and with even bigger stars, in a series of shows over the next four years. Then in 1975, ABC expanded the series to include a women's competition with tennis star Billie Jean King and gymnast Cathy Rigby and a bizarre celebrities' competition with fashion designer Oleg Cassini, music star Kenny Rogers, and actors Rob Reiner and Robert Duvall.[81]

Klein, of course, would capitalize on the series by dubbing Rotonda West the "Home of the Superstars" and use the tagline in pitches and promotional materials to sell lots—some 24,000 of them by 1975. However, the success of the Superstars television series belied the fact that Klein's Rotonda West development was imploding. The economy had gone south, prodigiously, while Cavanagh, under Klein's stewardship, had pocketed profits from lot sales at Rotonda West while defaulting on $10 million in debentures and $3.8 million in loans. Then, in February 1975, the New York Stock Exchange

Another scam was the futuristic Rotonda West development near Port Charlotte, whose lots, admitted the company in a 1979 settlement, were "of little value as investments and little use as homesites." The state banned the model pictured above for being deceptive, but salesmen used it anyway. Courtesy of the Paul A. Youngberg Collection, Charlotte County (Florida) Libraries and History Division.

suspended Cavanagh's stock, and the company filed for reorganization under Chapter 11 of the Federal Bankruptcy Act. The company's main creditors included the Vanderbilts and Chase Manhattan Mortgage, who foreclosed on pieces of Rotonda West and even its utilities to get their money back.[82]

Less than three years earlier, Cavanagh had been a $138 million publicly traded company with 1,500 employees. Now, it owed $78.8 million, while Rotonda West's 750 residents, living in 400 or so homes spread out over 18,000 acres, were literally and figuratively alone. Resident Richard Tanner, who built a home at Rotonda and bought three lots from Cavanagh for $10,000 apiece, took *St. Petersburg Times* journalist Dan Sewell on a tour of the development in 1981: "Tanner slows his car and points to a large, round concrete building that stands starkly on empty ground," wrote Sewell. "'That's a water storage tank that was never put into service,' he says. 'This was supposed to be the business section,' and that 'was going to be the commercial shopping center,' and 'up there was going to be a revolving restaurant. . . . We can keep going but it's more of the same,' and 'this is the good part I'm showing you now.'"[83]

Lacking roads or bridges or drainage, much of Rotonda West was inaccessible. It had one golf course, not seven as planned, while its unlit single-lane landing strip, which Klein had named the Rotonda International Airport, the "Airport of the Superstars," became a late-night drop-off spot for drug smugglers. In January 1974, for example, a twin-engine Lockhead Lodestar airplane carrying some 2,200 pounds of marijuana with a street value of $1 million crashed at Rotonda when its rear wheel hit a powerline. It careened into a field, landing safely, but, after a brief manhunt, its pilot, a thirty-three-year-old Connecticut resident named William Edward Klein (no relation to Joe Klein) was arrested. Meanwhile, a second pilot escaped, as did a large delivery truck that had been waiting for the plane to land.[84]

Residents at Rotonda had had enough, so in addition to mowing their own golf course themselves, they formed a property owners' association and retained legal counsel to, as one person put it, "force Cavanagh to meet its commitments."[85] Then, more than a dozen lot buyers filed suit against the company, charging "fraudulent misrepresentation," while in September 1975, the Federal Trade Commission (FTC) walloped Cavanagh with a thirty-eight-count complaint, charging that the lots it sold were "of little value as investments and little use as homesites."[86] The FTC complaint, as it were, was part of a larger federal push, begun in the 1970s, to reign in the biggest and most predatory of the land giants. In 1973, for example, the Office of Interstate Land Sales in the Department of Housing and Urban Development strengthened its registration requirements and cracked down on deceptive advertising by banning, among other things, artists' renderings of scenes not on offer, and in 1974 the FTC presented GAC, and in 1975 the Horizon and AMREP corporations, with point-by-point charges of unfair or deceptive trade practices that they could admit to and fix through a settlement or challenge (and face penalties to) in court.[87]

Cavanagh's charges were a litany of swamp-peddling perfidiousness. Among the lowlights: salesmen had sold lots while failing to disclose the fact that the land they were hawking was a fundamentally risky investment; had failed to give buyers property reports and/or offering statements as required by law; had touted the value of the property as increasing when they knew buyers would be unable to get what they paid for it or even sell it at all; and had lied by saying that Cavanagh would, if requested, buy the property back. Salesmen had also claimed that residents could access the Gulf by boat through Rotonda West's canal system when they couldn't; that the General Electric Company had planned and designed the community when all it had done was sell the developer appliances; and that Ed McMahon had taken an active part in the running of Rotonda West and had bought property there because of the soundness of the investment, when in fact he hadn't.[88]

Of course, Cavanagh denied the charges, but in late 1975, it had bigger fish to fry because, as part of its Chapter 11 reorganization process, the company's creditors insisted it give Rotonda West, or at least the management of Rotonda West, to a different and more capable developer—in this case the Deltona Corporation, which took over the project in May 1976. Deltona received a monthly fee for running Rotonda, plus a percentage of its overhead and additional compensation for marketing and construction, in all about $2 million a year for a period of three years. Although seemingly contradictory—with Deltona, a land giant, agreeing to improve property owned by a second land giant—the move made sense. With its stock price, its revenues, its land sales, its housing sales, and its condominium apartment sales down sharply in 1975, even before the Marco decision, Deltona would finish the year $5.1 million in the red. "It was like walking off a cliff," said Frank Mackle Jr. in 1976. "The sales were there one day and gone the next. . . . So we took a long hard look" and "reexamined where we [were] headed."[89]

By this point, middle brother Frank Jr. and son Frank III more or less directed Deltona, as Elliott had retired because of health problems in 1970 and Robert would retire in 1977. A civil engineer by training, Frank III was a thirty-two-year-old Notre Dame grad who'd been running Deltona's construction and development division since 1972. He was ambitious and capable, having read as a college freshman histories of the Ford, Du Pont, and Krupp families in order to glean insight, if he could, into how to run a dynastic company. But, as he entered the boardroom, the timing couldn't have been worse, for in June 1975 Frank III became a top-level executive at Deltona when the company was in decline. Then, less than a year later, the Corps of Engineers denied Deltona's Marco permits, and in May 1976, in what was a stopgap measure, Deltona took over Rotonda.

"We went to work sorting out the [Rotonda] mess," he wrote, at "the same time we were trying to dig our own way out of the impact of the Marco permit denials. We spent several months just figuring out where the project stood. How much cash was available for development? What was sold and unsold? What kind of contracts did they use? What had they told customers? What could be developed? When could [it] be developed?" And perhaps most important of all, in light of Deltona's Marco Island experience, "What were [its] environmental problems?" As Frank III explained: "The customers wanted answers. The banks wanted answers. The county wanted answers . . . [and] the press wanted answers," too. So, "after almost a year of study—and answering the questions the best we could—we [went public] with what we had."[90]

First off, Cavanagh's customers had signed some of the worst installment contracts in the business. Whereas other companies included the cost of improvements in the price of the lot, Cavanagh sold lots then billed customers for improvements: some $2,200 on an $11,000 lot, paid for in twenty semiannual installments, which could change if the price of construction changed. That meant that though Cavanagh was bankrupt, installment fees for some 24,000 lots continued to roll in. They also rolled in from Palm Beach County, where Rotonda East was a ghost development but from which Cavanagh collected and diverted some $2.4 million in fees. The monies from both Rotondas should have gone into escrow, but officials in Charlotte County found only $250,000 in Cavanagh's Rotonda account when millions more should have been there. However, when pressured by the Rotonda West property owners' association and its attorneys, not to mention the Division of Land Sales and the bankruptcy court, Cavanagh suddenly "found" the money, some $6 million of it, in Miami.[91]

Second, Cavanagh hadn't built much of anything, while the homes it had built were slapdash and crude. Author Jack Alexander described crooked lines, walls without framing, slabs poured and houses built with garages on the wrong side, and roofs *that hadn't been fastened to the walls*, bad in any case but lethal in a storm. "I was a building inspector up north," said one resident, "and when I saw the way they were building here at Rotonda, I had visions of roofs flying all over the county." Then there were design flaws and delays, such as when Illinois transplants Bob and June Melton moved to Rotonda West in 1972 but spent months in a condo because their home wasn't ready. One day June visited the site to find workmen installing a toilet bowl in the middle of a doorway—not next to the doorway, or across from the doorway in a bathroom where an architect would have put it, but out in the open where people would walk. Yet "when I complained," said June, "the guy said it was fine," that it was "low enough to step over."[92]

In terms of its improvements, due contractually by 31 December 1977, Rotonda lacked a complete and county-approved road system. Only the circle and a few residential roads had been paved, while the community's one semi-developed wedge, Oakland Hills, required five bridges, at a minimum, but had none. As everyone knew, Rotonda still had one golf course instead of seven, but no country clubs or clubhouses, and most of its canals, which were supposed to be navigable, hadn't been excavated yet. "Considering the relatively small progress made in building the first segment in the more than four years since the project began," lamented Joseph Tringali, president of the Rotonda West property owners' association, "how in the world can the enormous construction tasks for the remaining seven segments, plus completion of Oakland Hills, be accomplished in [time]?"[93]

The answer, said Frank III, was that it couldn't, not without modifying the original plan and extending the timetable from mid-1977 until January 1983. In other words, Deltona would finish the project, but in several key respects renege on what Cavanagh had promised. There'd be one additional golf course, for example, not six; powerlines would be above ground, not below ground; the development's bridges would be smaller; and Rotonda wouldn't have access to the Gulf. Nevertheless, Deltona promised to spend some $110 million in seven years, a startling amount that would come, apparently, from Rotonda's new lot buyers, if there were any; from its preexisting lot buyers, of which there were 24,000; and from Cavanagh itself, which reemerged from bankruptcy in December 1976. Both county officials and the Division of Land Sales approved the plan, but Rotonda's property owners were livid.

This wasn't what Cavanagh had promised. "[It's a] rip-off," exclaimed Tringali. Because if residents "are going to have a home on a green belt now instead of a golf course—with the approval of the government—then something is wrong." And "I'll be the first to admit that if they build seven golf courses, they're going to be bad golf courses. [But if Deltona doesn't build them,] will the man who purchased the lot get his money back? Or does he get to trade [his lot] for a lot on the golf course?"[94] With no say in the matter, and having been presented the plan as a *fait accompli*, in 1978 Rotonda's property owners filed a class-action lawsuit that eventually grew to over a thousand plaintiffs. Most joined the suit because in 1979 the FTC announced that though Cavanagh had accepted the commission's charges, signing a so-called consent agreement in which it pledged to clean up its sales practices, the company would not be required to give the money back.

The ruling was a win for Cavanagh, and for Klein, who by this point had shifted his focus from land sales to casinos. In June 1979, for example, using funds from Rotonda, Cavanagh purchased seventeen acres in Atlantic City's marina district for a 600-room hotel and casino complex it was planning. Meanwhile, Cavanagh had backed off improvements to the development, despite its seven-year, $110 million claim, then jettisoned Deltona in favor of a cheaper management company called Land Resources Corporation, the new owner of Lehigh Acres. Finally, in June 1980, Cavanagh sold Rotonda outright for $2.2 million, and title to an island nearby worth $8 million, to Rotonda Properties, Inc., a subsidiary of the same Land Resources firm that managed Rotonda.[95]

Thus, development-wise, Klein's reign of terror was over, at least in Rotonda, where residents spent the 1980s in a remote, desolate, poorly designed, semi-improved lot-sales vehicle that had failed—the only distinction was that unlike other land giant communities, Rotonda West looked an alien crop circle from the air. Even its new owners were confused, because buying Rotonda, explained William Futterer, the development's head of marketing in 1986, "was like buying a Rube Goldberg machine and receiving it in a crate in pieces, without assembly instructions and with some of the pieces missing. We found lot sales there to have been completely random. They were all over the place . . . with no visible plan. It was pretty obvious" that Klein "never intended to develop the property [at all]."[96]

CHAPTER SIX

# Generally Defective Communities

> It's proved to be a devastating problem for us. The configuration of the city was laid out with the wisdom of laying out a cemetery. They just divided it into a grid and sold off every piece they could.
>
> —BOB MINSKY, Former Mayor, Port St. Lucie, Florida

In Florida's Polk County, about an hour-and-a-half drive south of Orlando, is River Ranch, the one-time Gulf American and GAC subdivision on 43,000 acres of raw, unimprovable land. While GAC had gone bankrupt in the 1970s and creditors decided who owned what in a process that took four years, locals in Polk formed the "Hunt Club," a mud-bogging, swamp-buggying, ATV-driving, gun-toting, hog-hunting property owners' association that, while no one was looking, confiscated River Ranch. They simply fenced it in with barbed wire, set up an entry with armed guards and a caretaker, and refused access to any nonmembers who tried to go in, *even if they owned a lot*. The Hunt Club controlled roughly 10 percent of the land but sold its grazing rights for upkeep and claimed to be maintaining the property as the Rosens had intended: as a "vast and bountiful hunting preserve" where the owner of one piece of property, usually one and a quarter acres, could range freely and enjoy the property of everyone else.

"You buy [a parcel] in River Ranch and you hunt the whole place," said Hunt Club president Pete Edwards, a local truck driver who insisted that he and his compatriots, by enclosing the property, were simply "protecting the land." Better yet, "we're protecting *our* land," he said, "so we have to protect all of it."[1] That meant fences, guns, guards, and turning away anyone with even the slightest inclination of repurposing River Ranch, even retirees from Ohio who just wanted to see their lot. "Boy, I tell you, what a smelly-melly mess," said Dick Powell, a retired radio technician who in 1982 formed a rival land owners' association to battle the Hunt Club in court. "These guys have got tobacco dripping off their chin, a gun on their hip and a bulldog in the back of their buggy. As bad as you think it could be, that's what it is."[2]

In 1996, author Dennis Covington, whose father, a one-time steelworker and grocer, had purchased a lot at River Ranch following a sales dinner at a Holiday Inn in Alabama in 1965, decided to see the property for himself. It was his inheritance, after all; his father had left it to him. Covington drove to Polk County to check it out. He wrote about the experience in his 2004 book, *Redneck Riviera: Armadillos, Outlaws, and the Demise of the American Dream*.

> The gatekeeper, Mac McQueen, was snapping beans when we arrived. He had thick, callused hands, a shock of white hair, and a revolver strapped to his waist. "You're a long way from home," he said. "Don't they have enough deer up there in Alabama to suit you?"
>
> Mac's wife, Thelma, was more direct. "What are you doing down here?" she asked. She was standing at the counter underneath a calendar advertising deer stands and camouflage vests. She had narrow eyes and a crooked smile. The static from a CB radio crackled behind her.
>
> "I just wanted to show my nephew our land."
>
> "What section is it in?" Mac asked. I told him. "That land is worthless," he said without looking up from his beans. "You can't camp on it. You can't do anything on it, unless you want a thirty-aught-six going through you."[3]

But Covington ignored the threat, knowing by this point that bad publicity had forced the Hunt Club to allow people to enter who could prove they'd paid their property taxes by presenting the receipt. Covington showed them the receipt, then he and his nephew went in. What they found was a giant "camping" area with hundreds of trailers, cabins, A-frames, chanteys, lean-tos, and shacks, a veritable squatters' community with the haphazard ambience of a favela. Some owners used generators, if they had them, so extension cords stretched from structure to structure. There were Confederate flags everywhere—because this was Polk County, after all, where in 1967 a Klansman shot and crippled a black teenager in broad daylight for swimming at a whites-only beach—and trash heaps and discarded vehicles and dead washing machines and other appliances littered the ground. "There's more illegal building out there than Carter's got pills," said one observer. "Some live there full time."[4]

But they had no addresses or roads, let alone lot surveys. There was an ambivalent Polk County Sheriff's Department, some of whose own deputies belonged to the Hunt Club, and because Florida's Division of Land Sales viewed River Ranch as a property owners' dispute (and not a buyer-versus-developer dispute), it was therefore not under its purview, so River Ranch and its controversies have stretched into the present day. This was Covington's patrimony, his inheritance, which in many respects is Florida's inheritance, too: a legacy of 1950s-era land developers buying up property then subdividing the state, at times wantonly so, with little regard for planning, land use, transportation, infrastructure, water and sewer use, or the environment.[5] For raw-acreage subdivisions like River Ranch in Polk County, Golden Gate Estates in Collier County, and Cape Atlantic Estates and University Highlands in Volusia County, among many others, the problem has been what to do with the land. It's environmentally sensitive, so much so that private conservationist groups and public agencies at all three levels of government have attempted to buy it back.

In fact, the state of Florida considers River Ranch to be a key piece of its "Florida Forever" land acquisition program, which has as its goal the purchase and management of forests, wetlands, and fish and wildlife habitats deemed imperative to the "natural and cultural heritage" of the state. Each year, the program issues a priority list, and in 2018, "Bombing Range Ridge," a three-county area adjacent to the Avon Park Air Force Range, which includes River Ranch, made it to number two. The parcel, says the state, is a "high-need management area" because of its wide assortment of rare and endangered species, including black bears, scrub jays, red-cockaded woodpeckers, eastern indigo snakes, and burrowing owls.[6] Yet, though Bombing Range Ridge has been ranked by the state and listed for purchase for over twenty years, the Florida Forever program has found it impossible to purchase the land in bulk, at least not all of it, because River Ranch is a honeycomb of tens of thousands of lots whose owners have died or can't be located or who've given the property to multiple heirs. As one official explained: "The ownership pattern out there is a nightmare." And even if people own multiple lots, even hundreds of lots, "there's no one entity holding enough contiguous parcels that [an official can say,] 'Hey, I want to buy the 6,000 lots you own.'"[7]

Then, there's the Hunt Club, whose 4,000 or so members have banded together and hired legal representation to remove River Ranch from the state's priority list. "Any time the government buys property they fence it off and they control it," said former club president Pete Edwards. So, "we're going to fight the state. I own property in there [and] . . . bought it to take a gun in there and that's what I'm going to do." River Ranch "is about the last place for hunting in Florida that the state doesn't control" and "I bought this property with no intention to ever sell it."[8] With an assessed value of approximately $600 an acre in 2005, the lots that were available were cheap, certainly for the state, but as property boomed yet again in the early 2000s, speculators began buying them up to sell on eBay. They looked up the names of owners on tax rolls and sent thousands of form letters, making lowball offers and buying lots for as little as $200 apiece.

Next, they jacked up the price, ignoring (or not even knowing) the state's Division of Land Sales regulations requiring sellers to register, correctly advertise, and "disclose fully . . . all unusual and material circumstances" regarding the property and to refrain from implying, for example, that land values would increase or that roads or other infrastructure would be built. Yet in 2005, *Daytona Beach News-Journal* reporter Ron Hurtibise found University Highlands ads on eBay with descriptions that read "Here Comes the Neighborhood!" and "potential [value] of $50,000+ in the near future."[9] Hurtibise also discovered property sold not just to northerners but to foreigners and immigrants who had no idea what purchasing the land entailed. "Padamjeet Singh wouldn't be able to find the [University Highlands] property he purchased if he tried," wrote Hurtibise in 2005. "It's stuck on the fringes of a vast water recharge area—also known as a 'swamp'—just east of Deland. There are no roads to the property, just muddy trails," and "no one has surveyed it [yet]."

But Singh, a Maryland software engineer, had purchased two and half acres there from a South Daytona company called Davlar, Inc. for an eye-popping $7,900, when just two months earlier Davlar had bought the property for $1,000. "Singh thought he was making a sound investment," explained Hurtibise. "He figured he could build on his lot or make a profit by selling it in five to ten years. He didn't research the history of the property or seek advice from a real estate expert before buying," therefore when Hurtibise told him what it was worth, Singh moaned: "Oh, man."[10] Unfortunately, Davlar did a lot of business, selling some seventy-four lots in Cape Atlantic Estates and University Highlands at a profit of $4,000 apiece. But, because Davlar hadn't registered with the state or provided buyers with written reports of what they could or could not do with the land, it was forced in 2005 to refund over $300,000 and pay a $92,000 fine.[11]

That same year, James Kelly of Ormond Beach, the owner of a company called Volusia Wholesale Land and Property Investment, sold some $1.5 million in University Highlands lots to 190 mostly Haitian immigrants in and around Miami. Like Singh, they too thought they could build there—Kelly's representatives had told them as much, as did his advertisements—but after they'd bought the lots and discovered the truth, Kelly ignored their calls. "I trusted him and it's unjust," said a duped Delray Beach woman who paid some $16,000 for two waterlogged parcels. "For so many of these buyers, this was money that was earned from working in restaurants and farms and nurseries. This was honest, hard-earned money that [he] took." Apparently, the state agreed, because Kelly pled no contest to organized fraud, a felony, promised full restitution, and received fifteen years' probation.[12]

Imagine: investigations in 2004 by the State Attorney's Office into swamp peddling at University Highlands, a phenomenon journalist Morton Paulson first testified to before Congress in1964. "Paulson wrote a book about swamp peddling that had a chapter in it called 'By the Acre or the Gallon,'" explains Rob Walsh, a Volusia County official who buys up property at University Highlands and other "ghost subdivisions" for the "Volusia Forever" land program. "And he was right. Because what happened in the 1960s happened here again in 2005. That's when outta' nowhere people started showing up at the courthouse to purchase lots at auction whose taxes were overdue to sell on eBay. It was crazy, because prior to that, no one wanted them. The kids of the original owners just gave 'em up. They were escheated to us. That's when we thought, if we're really smart, let's try to buy up as many of these parcels as we can, because if not some huckster's going to say 'I can develop them!' and cheat someone else. Now, the eBay scams ended when the market crashed in 2009, and of course you're always going to have some bad apples. But we own enough pieces of property now that a big University Highlands development will never happen again."[13]

The same is true for the southern half of Golden Gate Estates, the onetime Gulf American ruse that Florida land-use planner John DeGrove called the "world's worst inappropriately platted subdivision."[14] Golden Gate, the planned community, and Golden Gate Estates, the nearby raw-acreage development, were built in the western Everglades, in the so-called Picayune Strand of the Big Cypress Watershed, a swampy area where seasonal rainwater from sloughs near Immokalee flowed south in a wide sheet to a dense mangrove coast and the incredibly productive Ten Thousand Islands estuary beyond. Environmentalists claim that the Picayune is among the most diverse ecosystems in America, but in the 1940s and 1950s, the J. C. Turner Lumber Company clear cut its old growth cypress and, not wanting to pay taxes on it anymore, sold some 61,500 acres to the Rosens for a song. Then, an additional purchase from the Collier family and two Miami companies in 1962 pushed the borders of Golden Gate Estates to an astonishing 112,000 acres, more than the cities of Miami and Orlando combined.

And because the land was on a gentle slope, the northern part was higher and drier typically, so Golden Gate Estates' first hardy settlers, and what rudimentary development it did have, went there. Gulf American (and GAC its successor) built hundreds of miles of roads in a grid pattern, but improvements beyond water and sewer lines for Golden Gate's core community, a mere five square miles in total, were not the company's concern. Therefore, the northern Estates grew gradually into a motley collection of trailers and ranch houses on two-and-a-half to five-acre tracts that served as a bedroom community to Naples. In fact, one observer called it the "anti-Naples," a place "where no homeowners associations are allowed and campers, boats, and RVs are welcomed on front lawns."[15] Yet, while northern Golden Gate—both the planned community and the Estates section—prospered and grew, from 1,230 residents in 1970 to 4,327 residents in 1980 and 14,148 residents in 1990, the southern Estates below "Alligator Alley," the Interstate 75 extension connecting Naples to Fort Lauderdale, died.

River Ranch notwithstanding, the "Estates South" section was arguably the weirdest place in Florida, where perhaps 100 homesteaders, without phone hookups or power, eked out a remote existence in scattered dwellings on 55,000 acres of land. "To reach the [subdivision]," wrote Florida author Bill Belleville in 1992, "I drive a circuitous route from the heart of Golden Gate to 'Everglades Boulevard' and travel across a lone I-75 overpass. I'm in a territory with no street or traffic lights, no water or sewer services and only random protection provided by police." Almost unbelievably, "in five hours of driving, I encounter one complete, neatly kept stucco ranch home, several wooden bungalows built from sheets of scrap lumber, a few hunting camps, and a half-dozen trailers." Someone had spray-painted street numbers on the asphalt, shot bullet holes through old Gulf American signs, and written graffiti that read: 'Pot Growing Here' and 'Cracker Rule!'" It was "deadly quiet," he continued, and "thinking only of the movie *Deliverance*," he left.[16]

That was probably a good idea, because the isolated Estates South section, or indeed all of Collier County abutting the Everglades, had seen a disconcerting number of corpses. In 1977, for example, the *News-Press* ran a story titled "Body Dump: The Mysterious 'Glades Holds Grisly Secrets," in which it quoted a Collier County Sheriff's investigator as saying that "at least 20 bodies" had been pulled from the swamp, including several in Estates South, in the seventeen years he had been there. And "that's probably 50 percent, or even less," he said, "of what's really out there."[17] Most were drug-related deaths, officials hypothesized, as Golden Gate's sheer remoteness, not to mention the fertility of its soil, made its southern and eastern reaches the perfect location for marijuana farms. There were hundreds of small plots, presumably, but in 1983, Collier County Sheriff's deputies confiscated some 4,000 plants from a single grow operation in Golden Gate, then in 1988 seized 2,700 more.[18]

Gulf American's Golden Gate Estates development was the quintessence of swamp peddling. Sold as "investment acreage," the Estates property south of Interstate 75, all 55,000 acres of it, was pitched to buyers as a future suburb when, without multibillion-dollar improvements, it could never be built. Courtesy of the National Archives (412-DA-6111).

What with its wide, remote boulevards, the Estates South section was a favorite landing strip of drug smugglers who crashed or were caught in police stings in 1975, 1976, 1978, 1980, 1981, 1983, and 1984. The payloads were staggering: 1,000 pounds of marijuana packed into bales in a twin-engine plane from South America; 900 pounds from Jamaica; and Collier County's biggest seizure of all, 876 pounds of cocaine worth a reported $175 million from the Bahamas. In a scene reminiscent of the TV series *Narcos*, at 1 A.M. on the night of 24 May 1984, a group of five individuals in a Cadillac, a pickup truck, and one other vehicle sat on a lonely stretch of Everglades Boulevard ten miles south of the interstate. They were waiting for a delivery from a twin-engine Piper Aztec, but U.S. Customs officials had tracked the plane on radar and had sent a Black Hawk helicopter to intercept it.

The Black Hawk found the plane being unloaded, and the pilot and people waiting for it scattered into the bush. Two of them jumped into the pickup and blazed through Estates South in the dark with the Black Hawk in pursuit. They eventually ditched the truck, fleeing on foot, but everyone who had been at the scene escaped—except, perhaps, for a twenty-six-year-old Colombian national found filthy and wet miles away at a payphone.[19] Equally as dramatic was the failed apprehension a year later of drug smugglers in a Beachcraft Queenair loaded with marijuana that Customs officers had tracked to Golden Gate, but whose pilot had noticed a spy plane in the vicinity and decided to jettison his cargo by dropping perhaps two dozen fifty-pound bales into the swampland below. The pilot landed nearby, on an Indian reservation in the Everglades, but when tribal police searched the aircraft they only found residue. "There was marijuana on that plane at some time," said one officer, "but where it is now, who knows?"[20]

On occasion, Collier County Sheriff's deputies used tips and airborne reconnaissance to conduct sweeps of Estates South to catch growers in the act. However, during one such sweep in 1983, deputies encountered two unarmed Nicaraguans in combat fatigues walking along 69th Avenue SE—a pockmarked road miles from anywhere. When asked what they were doing, the men freely admitted they were members of the FDN, the Nicaraguan Democratic Force, an anti-communist, pro-Contra, CIA-affiliated commando unit that trained, of all places, in Golden Gate Estates. The FDN had a camp with a mess hall and a shooting range and put up posters that read: "With God and Patriotism We Will Destroy Communism!" and "The Pope is with Us! Christ is the Liberator!" The camp's owner was Geraldo Silva of Miami, a member of Brigade 2506, the ill-starred Bay of Pigs invasion force from 1961. Nevertheless, the sheriff's department found that neither Silva nor the FDN had broken the law, but Collier County zoning inspectors, in a Sandinista-like countermeasure, shut them down.[21]

Clearly, the Estates South section was a wild place, but 279 miles of roads, 251 culverts, forty-eight miles of drainage canals, and eight weirs had devastated the land environmentally. "These were all swamplands," said Ananta Nath, an engineer with the South Florida Water Management District, "and the only way they could develop the land was by draining the whole landscape. Before the area was developed, water used to flow slowly like a sheet flow to the Ten Thousand Islands. Now the roads block it and the canals are bringing it like a river."[22] Indeed, according to one 1990 estimate, 363 million gallons of freshwater flowed through the development's canal system and into the Gulf every day. The result was that Golden Gate's water table dropped two to four feet, which dried the scrub brush significantly and led to forest fires, including a series of huge fires in January and February 1985 that burned over 30,000 acres, injured twelve firefighters, and killed a forest ranger.

To drain Golden Gate Estates, which sat at the far western edge of the Everglades, Gulf American built eight weirs, 251 culverts, and forty-eight miles of canals that dried the surrounding lots but devastated the land environmentally. Courtesy of the U.S. Army Corps of Engineers/Flickr.

"What happens when you dig these canals," explained one observer, is that "you drain the land. You're messing with Mother Nature. It's made to work a certain way. [But when they built these subdivisions] they drained a lot of land that's supposed to be wet."[23] What is more, frequent wild fires destroyed the underbrush, killing animal habitats and allowing harmful and invasive plant species, such as the forest-producing melaleuca tree, to flourish and in some areas replace what had been a river of grass. And, each wet season, millions of gallons of freshwater flowing south into Faka Union Bay upset the tenuous freshwater/saltwater balance of the Ten Thousand Islands chain, lowering fish and invertebrate numbers and killing seagrass beds. Then, during the dry season, when outflows were low, saltwater moved up the Faka Union Canal and into the groundwater, which decimated agriculture and posed an existential threat to the region's water supply.[24]

"Some people call it a horror story," wrote environmental reporter Betty Price in 1977. "Others liken it to a desert, or a devastation scene from a science fiction movie."[25] But, whatever it was, the poorly planned, environmentally decrepit Estates South section needed a comprehensive restoration program to sort it out. Therefore, in 1985, the state of Florida added the failed subdivision to its Conservation and Recreation Lands (CARL) program, a government acquisition fund similar to today's Florida Forever program, in which officials attempted to buy Estates South, all 55,247 acres of it, in its entirety. The intent was to restore its "pre-drainage hydrology and ecology" by removing its road and bridge network and filling in its canals.[26] But to do that, some 19,621 lot owners would have to be found, then sell their property for thousands less than they paid for it, sometimes for just $300 an acre. Most owners balked; some went to the newspapers and formed not one but two protective associations. By 1990, a paltry 1,300 acres had been sold.

And all the while, Avatar Properties, Inc., a subsidiary of Avatar Holdings, the company that succeeded GAC Properties, continued to sell lots. In 1987, for example, the *Miami Herald* reported that Avatar had been outbidding the state to buy acreage from long-suffering northern owners, then selling the property at a markup to the Latin community of Miami. With ads on Spanish-language television and in newspapers, Avatar fooled people, such as a thirty-two-year-old upholsterer named Alvaro Moreno, into thinking they were buying *fincas*, or ranches, that could be transformed into Latin American–style estates. Some lots were in northern Golden Gate Estates and thus were buildable, but "scores of recent sales," reported the *Herald*, were in the south. "They really convinced me this was a good investment," said Moreno, who in August 1987 paid an eye-popping $16,000 with $5,000 down and a ten-year mortgage on a 2.3-acre lot. However, in October, he and his wife returned to the property. It was so remote, and there were so many mosquitos, explained Moreno, "My wife didn't even want to get out of the car."[27]

Meanwhile, a second scam, the state alleged, involved a for-profit company with a nonprofit-sounding name, the Golden Gate Landowners Association, Inc. (GGLA), which for a $35 annual fee promised to "protect the interests of absentee landowners" by serving as a "bargaining agent" with the state and negotiating a purchase price "two or three times what [owners would] receive by acting alone."[28] The problem, however, was that the GGLA wasn't a bargaining agent at all, because the state didn't bargain. By law, it simply offered owners the average price of two independent appraisals and that was it. But northerners didn't know that, so the GGLA and its president, a convicted drug smuggler named Larry Fitzpatrick, tricked them into paying for a service that, in effect, was little more than a newsletter whose 4,200 subscribers paid GGLA some $150,000 a year. Golden Gate owners "have already proved their gullibility by buying land sight unseen," wrote one journalist. Now they're being "hit up by crooked surveyors, appraisers, attorneys, and numerous other bunko artists" who say that "they just want to help."[29]

Spurred by owners' complaints, in 1985 the Economic Crimes Unit of the Florida Attorney General's Office ordered Fitzpatrick and two partners to send letters to property owners "correcting misleading representations" and to refund the $35 fee to any member who asked for it.[30] The partners did as they were told, but continued their GGLA information service—a four-page newsletter mailed to members every ten weeks—while "helping" members sell lots to the state (at preset prices) for a commission of 15 percent. Fitzpatrick "sends letters out that he's representing you, blah, blah, blah," said a Fort Lauderdale dry cleaner who owned property in Estates South. "He's not representing anybody [but himself]."[31] Indeed, Fitzgerald would admit he was in it "for the money," so much so that he and his partners formed similar for-profit associations for owners of Remuda Ranch, Flagler Estates, Viking Estates, Imperial Acres, Sunrise Acres, Orlando Pines, North Port, Port Charlotte, and Deltona.

The goal was to collect membership fees from as many aggrieved landowners as possible, from those in Estates South wanting higher appraisals from the CARL program, for example, or those in planned communities such as North Port wishing to sue a developer over missing amenities. But the main strategy that these for-profit landowners associations employed was the class-action lawsuit—or rather, the promise of a lawsuit—which a wily character named Mark Binstein was adept at getting people excited about. Originally from New Jersey, Binstein had been a securities dealer in the 1960s, then later coach and general manager of the defunct Pittsburgh Condors franchise of the American Basketball Association, and he once spent fifteen days in jail for going into the stands and punching a Miami Floridians fan who had been heckling him.[32] The team went bankrupt in 1972, so Binstein rebranded himself as a "land fraud investigator" by forming landowners associations for failed developments in Colorado, New Mexico, and Florida and asking members to put up 8 percent of their purchase prices to pay for lawsuits.

He toured the country, giving impassioned speeches to jilted buyers in Holiday Inn ballrooms and community centers, where he promised people 80 to 100 percent refunds if they joined his associations and sued. Yet Binstein wasn't a lawyer; he was an "investigator" who covered his $2,500-a-week salary and expenses from money he had raised to hire lawyers. That was the business, and that, critics said, was the con. "I'm paid and paid very well for what I do," he explained. "I'm not Ralph Nader. I'm not a crusader. As a lay person, I can't share in a legal fee or recovery. [So] I'm paid out of the war chest from the associations."[33] Paid was an understatement: In 1978, for example, Binstein raised $220,000 for his Pueblo West Landowners Action Association, which was suing a company called McCulloch Properties for misrepresenting land it had sold in Colorado. Binstein then gave an attorney $30,000 to represent the group but spent $190,000 on himself.[34]

He collected $381,000 from owners of AMREP properties in New Mexico, pocketing a $145,000 salary plus $16,000 in expenses as the AMREP landowners association's only employee, but paid just $103,000 to lawyers representing the group.[35] Naturally, the companies targeted by Binstein complained, calling him a conman and crook, while several, including McCulloch Properties and the Great Western Cities Corporation, owner of the Colorado City development, filed their own lawsuits, claiming Binstein's "solicitation-for-profit enterprise" used purposely false information to inflame litigants in what could be only described as a ruse.[36] By 1980, Binstein faced lawsuits and/or administrative actions in Texas, Minnesota, Wisconsin, Indiana, and Illinois, while a judge in Indianapolis barred him from soliciting clients for his Pueblo West association altogether. "A Supreme Court Justice once said he couldn't define obscenity, but he knew it when he saw it," said Judge S. Hugh Dillin. "Well, this is an obscenity if I've ever seen one. [Because] this man, in the name of saving your property, is ripping you off."[37]

Undeterred, in the 1980s, Binstein shifted his focus to Florida, where he formed landowners associations for North Port and Rotonda and recruited some 3,800 members. At upwards of $600 apiece, Binstein made millions investigating Ed McMahon and Joe Klein, among others, which brought publicity but little if anything to those who purchased lots.[38] Eventually, though, karma prevailed. In 1994, Binstein asked members of the North Port Out of State Landowners Association if they'd care to contribute $100 for a case he was handling—his own. That year, the Internal Revenue Service accused Binstein of failing to report some $250,000 in income and of claiming $180,000 in bogus deductions. He said the charges were financially motivated, the result of lawsuits and investigations that, "thanks to me," had caused "thousands of people [to stop] paying their mortgages."[39] However, in 1995, a federal jury in New Jersey disagreed, deliberating for just ninety minutes before convicting the land fraud investigator, who had held over 300 landowners' meetings in seventeen states and Canada that had earned him over $2.5 million, of fraud.[40]

Now, back to Golden Gate, where by the early 1990s GGLA president Larry Fitzpatrick had overcome his run-in with the state to create, almost by accident, a legitimate landowners association. For in addition to writing a newsletter and making money off it, Fitzpatrick realized that the state's appraisal process was unfair, even actionable, because when it placed Estates South on the CARL program's to-be-acquired list, the state had asked the county to restrict permitting there as part of a state-mandated growth-management plan.[41] This had unlawfully suppressed prices, so Fitzpatrick contacted Craig Boda, a hard-driving, drug-using, soon-to-be-disbarred Flagler Beach attorney, with a plan: If Boda would represent the GGLA and its 4,000 or so members in a lawsuit with the state, they would split Boda's recovery fees fifty-fifty.[42] But fee-sharing between lawyers and nonlawyers was illegal, so much so that even Boda, whom authorities would catch laundering money for a drug dealer client through his office's escrow account, wouldn't do it.

Perhaps it was impractical, what with so much money involved, so Fitzpatrick devised a different plan: He would charge members a $195 investigative fee à la Mark Binstein, then have them pay him a commission for their lot sale if Boda settled with the state. And, though Fitzpatrick lived five hours away in Daytona Beach, he obtained a brokers' license and opened a real estate company to do it.[43] With Boda as its counsel and Fitzpatrick its president, the GGLA would sue the state in 1988 and 1992, but as it waited for a settlement, some 60 percent of all Estates South landowners sold out. Nevertheless, the GGLA wouldn't budge. The state "[sees] us as a bunch of profiteers out to make a buck and not giving a damn about the Everglades," said Boda. "That's just not so. We're fighting for an appraisal process that's not tainted. Everybody realizes that [Estates South] shouldn't have been platted . . . in the first place."[44]

What broke the impasse was the announcement in June 1997 by Vice President Al Gore at a speech in the Everglades that the federal government and the state of Florida would each contribute $25 million to purchase lots at Estates South for their inclusion in a newly created state park called the Picayune Strand. "Thirty years ago when we didn't know much about the environment," said Gore, on a ninety-degree day with mosquitos buzzing about him, "south Golden Gate Estates was intended to be part of the largest subdivision in the world. [That's when Gulf American] brought in people from the north in the wintertime to show them the plots—in the dry season."[45] The Gore announcement was a big one, a "breakthrough" said observers, which allowed the state to offer the GGLA higher appraisal prices—an increase in its original offer of between $300 to $2,000 an acre to $1,000 to $3,000 an acre—for a payout of $15.8 million.[46] The GGLA accepted, as did numerous nonmembers who also took the deal, but with 11,000 acres still to be purchased, in 2001 then governor Jeb Bush approved (what amounted to) 1,800 forced buyouts through eminent domain.[47]

By 2005, the state had purchased nearly all of the Estates South tract, but two landowners were left: Florida's Miccosukee Indian Tribe and an elderly, disabled Naval veteran named Jesse James Hardy. On the surface, the Miccosukees would seem to have been the most unlikely of installment buyers because, in fact, they weren't. In 1997, in the midst of the state's buyout program, they slipped into the picture and purchased 800 acres in two large parcels southeast of the development's road grid from an Illinois phosphate company for $437,500, which was more than the state had been offering. Tribe attorney Clay Reiner insisted that the Miccosukees had only meant to use the land for gathering plants for medicine and palm fronds for the thatched roofs of their chickee huts, but, tellingly enough, they'd recently been fighting with the state over the wanton pumping of billions of gallons of runoff that flowed south from ranches and farms and through their reservation, sixty miles to the east of Golden Gate, in the Everglades.[48]

The tribe wanted the pumps turned off, not to mention tighter pollution restrictions targeting the sugar industry in particular, so it filed a federal lawsuit and threatened to use tribal sovereignty to enforce is own runoff restrictions to do it. That's when it purchased the Estates South property, which had no real role in the debate but whose acquisition by the tribe threatened to derail the state's entire reclamation program because, if even one owner remained, the state couldn't deny the owner access to the property by purposely flooding the land. It was a brilliant strategy, for the Miccosukees were pressuring the state in one dispute by injecting themselves into another, but this was the same tribe that, while publicly defending the Everglades from runoff, had asked the Shell Oil Company to drill exploratory oil wells on its land.[49] Therefore, the state, in attempting to finish its Estates South acquisition program, had little patience for the Miccosukees.

"They're in the path of major restoration projects that are vital for the long-term survival of the Everglades," said Governor Bush. "[So you have to ask,] are these tribal lands or are they lands that have been purchased by the tribe? There's a difference."[50] The difference, of course, was eminent domain, which Florida used to acquire the property in 2005 for an extremely generous $2.2 million. The tribe complained but didn't pursue the matter, having made a tidy $1.8 million profit on land it owned for just eight years. Meanwhile, the state's other holdout was Jesse James Hardy, who purchased his 160-acre parcel in 1976 for $60,000 and who lived in Estates South like a homesteader with the niece of an old girlfriend and her nine-year-old son. Hardy considered the boy to be his son, the woman his family, and planned to leave the property and a small quarry he ran there to the two of them when he died. That, he claimed, was non-negotiable, no matter what the government offered.

"It's my home and it's where I run my business and where I raise my family," he said. So "why should I leave just so the government can conduct its billion-dollar science project? I'm 69, and having a comfortable place to live where I am happy means more to me than money. There ain't no price tag on giving up happiness."[51] For a time, Hardy became a cause célèbre among property-rights advocates nationwide, who created a website for him and traveled to Tallahassee in mass for a raucous meeting with the Board of Trustees of the Internal Improvement Trust Fund and who also advocated for him with a song, "The Ballad of Jesse Hardy," written by a retired Navy Seal named Larry Bailey. Eventually, though, when the state upped its offer from just under $712,000 to $4.95 million and filed for eminent domain, in 2005 Hardy grudgingly agreed to go. "I'm not going to shoot nobody," he explained. "I'm not going to fight nobody," and "I'll get the hell out of here. But I got took." Perhaps in his heart of hearts Hardy believed that, but just two months after settling with the state, he purchased a $750,000 home in Naples and incorporated, of all things, the Jesse James Hardy Realty company to sell property in the Panhandle.[52]

FLASHBACK TO 1959, when the Mackle brothers/General Development partnership was in full swing: Port Charlotte was booming, sales-wise, so the Mackles planned what they called a second community "core" nine miles north of Port Charlotte in Sarasota County. The idea was that Port Charlotte and "North Port Charlotte," as the new section was known, would both bookend a giant community the width and breadth of the Frizzell Ranch that would include a truly remarkable 200,000 lots. Therefore, to sell the northern section, the Mackles built eleven model homes and added some paved streets with sidewalks and a small utilities system—the usual lot-selling gig.[53] However, for properties beyond the core—that is, the bulk of North Port Charlotte, or "North Port" as it was known—the brothers eliminated the amenities altogether. Roads, for example, were of poorer quality and lacked sidewalks, while each home's utility system was a well and a septic tank.

But, when the Mackles submitted the plans to Sarasota County, officials there said no. They wanted to control growth, to ensure that an isolated development at the far end of the county wouldn't drain services or that, when it rained, tens of thousands of septic tanks wouldn't overflow.[54] In neighboring Manatee County, five days of thunderstorms in March 1959 saturated drain fields, pushing sewage and effluent into yards and driveways and filling its ditches with excrement. "It's a health hazard," bemoaned one official, and "the only way the bacteria around [these] tanks will disappear is for the sun to shine on it. [But for that you need] eight-inch sewer lines and a disposal plant"—a multimillion-dollar infrastructure the Mackles had no desire to build.[55] Thus, facing an impasse, the brothers pulled a fast one. Rather than build the infrastructure, they appealed to Tallahassee and a pro-development legislature that allowed North Port Charlotte to incorporate as a city, the plan being that it and its twenty-one residents, almost all of whom were GDC employees, would be friendly to the Mackles.

The move was a work-around to (what for the time were) fairly advanced county zoning laws and planning practices in favor of a company town, which North Port Charlotte became, at the behest of the Mackles, in June 1959.[56] The vote was twenty to one, and overseeing the new city "government" was a mayor and four commissioners who worked for the company and, as politicians, did whatever the company asked. For instance, as of January 1960, North Port Charlotte had no city-owned buildings whatsoever, no fire department or police force, but took it upon itself, at GDC's urging, to formally annex from the county some 16,640 acres of GDC-owned land. That exempted the company from county regulations covering everything from zoning and drainage to utilities delivery, lot-size minimums, septic tanks, and roads. Naturally, county officials were infuriated. "The whole thing is a joke," said an angry planning director, William Vines, and it's "not a funny joke either. There is absolutely no reason for the incorporation of this area except to avoid county regulations. . . . [Now] all of the normal rules that apply to developers are out the window."[57]

But that's just it: In the 1950s and 1960s, the land giants had no rules, at least planning-wise. They built what they wanted to build and chose how they wanted to build it because they operated in rural sections of the state, in counties with small governments eager for the increased tax base—Sarasota notwithstanding—and where zoning regulations and planning boards were unknown. In fact, prior to the 1970s, only municipalities, meaning cities and towns, had the blanket legal authority, under Florida law, to insist that such-and-such an area be zoned commercial, for example, or that buildings be no more than a certain height or set certain distances from each other, or that neighborhoods have sidewalks or build-to lines or a certain quality of road. However, the counties were a different story. Some of the larger or more progressive ones (such as Sarasota) had asked for and received special acts of legislature allowing them to plan and zone, but dozens of Florida counties had no regulations whatsoever—no planning, no zoning, not even building codes.[58]

In sparsely populated Charlotte County, for instance, by late 1961 GDC had built some 4,000 homes, a shopping center, a bank, a post office, a yacht club, a bowling alley, six churches, a thirty-unit hotel, and a parochial school *without ever needing a permit*. The company filed plats, of course, and paid registration fees, but no one cared what GDC did, as long as it did it. "Those backhoes, and those sales pitches, and those resort hotel rooms created jobs," said Wayne Daltry, a retired official with the Southwest Florida Regional Planning Council. So "there was no immediate demand or much interest in regulatory controls. Heck, the counties weren't even authorized to regulate [developers] without a special act of the legislature."[59]

In neighboring Lee County, where land giant developers began building communities at Lehigh Acres (1955) and Cape Coral (1957), a county planning commission required plats, a drainage plan, and "performance bonds" guaranteeing that, in the event that their business failed, developers would still provide customers with roads and other improvements they had promised. But "your field is unlimited," commission chairman Wilson Pigott assured developers in 1959. "[We just want things to be] heading in the right direction."[60] Nevertheless, the county's own lawyer advised the commission that without an act of legislature, literally "none of the county's laws pertaining to zoning" would stand up in court, and "that only the voluntary cooperation of developers" had "kept subdivision development in any order."[61]

But, it was the wrong order, ultimately, because the Rosens, the Kleins, and the Ratners of the world weren't building communities, per se. They were building lots, hundreds of thousands of lots, which salesmen sold to northerners as soon as the company could plat them. Just look at Google Maps: Port Charlotte, Cape Coral, North Port, Port St. Lucie, Palm Bay, Citrus Springs, Rotonda, Deltona, Palm Coast, Golden Gate, Spring Hill—they have no downtowns, small numbers of sidewalks, and relatively few parks or industrial zones, while banks, grocery stores, restaurants, box stores, and strip malls line ribbon-like business districts on either side of a road. "It's proved to be a devastating problem for us," said former Port St. Lucie mayor Bob Minsky. "The configuration of the city was laid out with the wisdom of laying out a cemetery. They just divided it into a grid and sold off every piece they could."[62]

Author Paul Reyes put it a different way: "The idea from the beginning . . . was to sell lots," he wrote, "nothing more. No forethought was given to the possibility that someone might actually want to live there, no consideration given to streets or schools or sewers. . . . Selling, not building, was how these developers measured their success."[63] And they were a success, because today, Florida has some 2,600 subdivisions with what experts call "obsolete" or "antiquated" plats. (Some, not so charitably, call them "dinosaur" plats.) That's 2.1 million lots, laid out before planning or when planning was in its infancy, that have haunted Florida by skewing and/or and delimiting its development to this day. "Conventional wisdom on what constitutes optimal neighborhood design is constantly being reevaluated," wrote a state legislative committee in 2004, "but there appears to be [a] consensus that these old platted subdivisions do not carry traits that are conducive to providing a high quality of life."[64]

For one, the majority of all lots in antiquated subdivisions were 80 feet wide by 125 feet long, or 0.23 acres in size. In Lehigh Acres many were 0.25 acres in size and deed restricted from when the Lehigh Acres Development Corporation first platted them as "single-family residential" in the 1950s. That meant one lot with one house, or more commonly, a vacant lot, 135,000 times. Meanwhile, the company had set aside just 2 percent of the community for businesses, a sufficient amount in 1970, when Lehigh Acres had just 4,000 people.[65] But fifty years later, with a population in the 120,000 range and growing, Lehigh is bursting at the seams. It has plenty of land, ironically, but not enough large parcels for business districts or industries. Therefore, companies can't move there; jobs are low paying or nonexistent or tied to the building trades, which are cyclical, while a series of Florida laws, passed in 2006, prohibit government agencies from using eminent domain to acquire privately owned residential properties for economic redevelopment.

Thus, for communities like Lehigh, assembling one large parcel from dozens of smaller parcels is incredibly hard, for ownership is fragmented and rezoning is time-consuming and difficult while many owners refuse to sell. But such a lack of commercial space isn't just a jobs issue; it's a tax issue, too, because as Cape Coral's economic development director Mike Jackson explained in 2005, cities derive on average 25 to 30 percent of their property tax income from commercial properties. But Cape Coral was at 8 percent, which meant "residents pay a disproportionate share of the taxes." After all, said Jackson, "businesses use [fewer] services and pay greater taxes because of their size than residents do. [They] don't strain schools with additional students, don't ask for parks, and don't use police or emergency medical services as often as residents [do, either]."[66] And, though Cape officials have worked to attract businesses by helping to create commercial property, and have even held "Plat-a-Palooza" forums to solicit ideas from the public, explosive population growth and a new round of building in the city's outer districts have kept its commercial-to-residential ratio the same.[67]

Then there's the geographic size of the communities—they're huge. Cape Coral, for example, encompasses 120 square miles, which, says Cape resident and noted planning expert Max Forgey, is "like taking the city of San Francisco, which is 47 square miles, and putting it on the south end, then taking the city of Boston, which is 48 square miles, and putting it on the north end, and having enough left over for Bloomington, Indiana, in between."[68] But here's the rub: The Rosens sold lots in Cape Coral and customers built houses there at random, because no one told them not to. The result has been that basic services, such as water or sewer or trash, have been wildly expensive to provide. In the late 1990s, for example, in Port Charlotte, garbage trucks could service 1,200 homes a day. But, in isolated areas workers could service just 300 homes a day. "You just drive forever between stops," said the head of a local sanitation company, and "it's not uncommon to find one house on five or six streets."[69]

Firefighting was the worst. In the sprawling development of Palm Coast, near Daytona Beach, in the 1970s, the ITT Levitt Development Corporation—a company created by the acquisition in 1968 of famed developer Levitt & Sons by the manufacturing conglomerate International Telephone and Telegraph—built lots and roads on 42,000 acres of pine forest for a planned community of 225,000 people. But, by 1985, it had less than 8,000 people, who lived mostly in neighborhoods close to the coast, but also in scattered homes in a labyrinth of tree-lined streets, some ending in cul-de-sacs, stretching for miles. A house abutted the woods, followed by another house and more woods, but neither ITT nor Flagler County had considered the dangers of wildfires when Palm Coast was built. That was unfortunate, catastrophic even, because Palm Coast was situated in what fire experts call a "wildland-urban interface," an area where housing developments meet and intermix with forestland and where wildfires are common. In May 1985, for instance, high winds and unseasonably dry weather fueled a series fires in at least twenty Florida counties, including Flagler, where Palm Coast's two trucks and twenty-seven volunteer firemen were overwhelmed.

"We tried to get in front of it, but the flames went right by us," said an exhausted Palm Coast fire chief, John Brink. "It was too big, too wide, and too tremendous."[70] Instead, as flames moved from lot to lot and passed through treetops overhead, firemen risked their lives by racing up and down streets and into cul-de-sacs—which now were death traps—to warn residents to get out. Thankfully, Palm Coast's 1985 fire didn't kill anyone, but it destroyed 131 houses, while burning 22,000 acres and killing 1.2 million trees.[71] Things could have been better if officials had adopted so-called defensible-space standards for its residential codes, including, most notably, a clearing requirement of at least thirty feet between houses and forest. But, because ITT had platted so many 80′ × 125′ lots, clearings were ten feet, at best, sometimes seven, so there was nothing officials could do.

"I am enforcing [defensible-space standards] when there is a new subdivision coming in," said an exasperated Flagler County fire marshal in 1998, after a second huge fire spewed "ash the size of cars" and jumped a highway and an interstate to destroy forty additional homes, but "you can't change anything [in Palm Coast] because you can't change lot sizes."[72] You could, though, prohibit flammable wooden street markers, which ITT had thought rustic; require residents to clear underbrush or remove combustible pine trees from their lots; or conduct controlled burns. The problem, however, was that Florida law mandated that only the state's Forestry Division could conduct burns on private property without the owners' consent; therefore, to hold such a burn, counties like Flagler had to track down and convince thousands of lot owners to do it. That wasn't practical, or even possible, so Forestry did the burns, but officials there hated to set fires in woods "dotted" with houses, so handled, in the thirteen years between the two fires, less than 400 lots a year.[73]

Next, water to fight fires was also a problem because, in the beginning, land giant developers typically provided residents with few, if any, utilities. Water lines were expensive; so were sewer plants. So in the 1950s they confined their utilities to a central core, adding them retroactively in most cases, street by street, and charging residents for the job. However, houses in remote neighborhoods got nothing. Instead, they used water wells and septic tanks, since running eight-to-ten-inch pipes eight miles into the bush, to Lehigh Acre's sparsely populated western section, for example, was prohibitively expensive. Thus, as of 2006, roughly three-quarters of the 143-square-mile Lehigh Acres Fire Control and Rescue District lacked fire hydrants, which forced crews to transport water in tenders.[74] "For us [that's] pretty much just the normal response," said a fireman in Golden Gate, where in 2010, experts estimated that bringing water and sewer lines to a five-square-mile section of the northern Estates area would cost $145 million or, for landowners, $107,000 a parcel.[75]

Equally expensive were the roads, for land giant developers built thousands of miles of roads as quickly and cheaply as they could. But instead of using lime rock or coquina stone as a base, then covering that with one and a quarter inch of asphalt, which is what cities and counties would do, the developers used a six-inch-deep soil-cement base with a one-inch asphalt covering, mixing Portland cement and local soils on-site in giant "single-pass soil stabilization machines" that could manufacture 450 to 500 tons of road base per hour.[76] The soil-cement method was cheap—one 1975 estimate put it at $7 a running foot compared to $16 to $17 a foot for other methods—while the use of soils sourced locally (and not brought in from elsewhere and purchased) allowed developers to build entire blocks of streets in a single day.[77] However, unlike a lime rock or a coquina stone base, a soil-cement base was exceptionally rigid. Therefore, when it cracked, the asphalt above it cracked, in a process known as "reflection," leading within a few years to what Port St. Lucie engineer Walter England called "a worn-out road."[78]

The biggest builder of "worn-out roads" was the General Development Corporation, which used a soil-cement base for upwards of 4,000 miles of roads in its various Florida developments, before dumping the thoroughfares, and their upkeep, on local governments. "We didn't have to accept them," says Tom Frame, a former zoning director and administrator in Charlotte County in the 1970s and 1980s, "but those roads were public rights-of-way. So at some point, in theory, GDC could have abandoned them. And remember, the people who lived on those roads were paying taxes to the county, so we also faced political pressure to fix them. But before we'd accept a road and take responsibility for it, we'd insist on certain minimum standards, a better road base or a thicker black top, for example. Then, GDC would do it, or we'd get money from GDC and do it ourselves."[79]

However, that was the best-case scenario, when adequate planning measures were implemented in the 1970s. In the 1960s, though, counties including Charlotte sometimes accepted roads prematurely, without changes, or companies refused to make them. In 1961, for example, after a protracted dispute with St. Lucie County over the efficacy of building hundreds of miles of soil-cement roads in its Port St. Lucie development, GDC pulled a second "North Port," so to speak, by moving its employees into a handful of houses there, holding a referendum, and seceding.[80] Although the new Port St. Lucie city had no municipal buildings or police force, its very first act of government was the implementation of a GDC-friendly building code. No doubt, county officials had been right to be wary because at the time, soil-cement roads were experimental. There just weren't many in Florida, except, perhaps, for a half-mile "test strip" in Sarasota County and a three-and-a-half-mile stretch in St. Johns County, near St. Augustine, which Port St. Lucie officials inspected in 1959.[81]

The roads were passable, they found, but without better data or more insight as to how soil-cement performed over time, they "had no yardstick," as one commissioner put it, who then asked, "What would happen in ten years [time] if the roads went bad and the county [had to perform] maintenance on them?" Because, "once we accept these roads, they're our baby."[82] Charlotte County officials would learn that lesson the hard way, when in 1964 they accepted for maintenance all roads platted in subdivisions in the county before 1960, then in 1965, added without conditions the majority of the roads then existent in Port Charlotte.[83] There were hundreds of them, and some were so poor, so gashed and pockmarked, that they had weeds growing through them, a kind of "grassphalt," joked locals, which county maintenance crews then had to mow. "When these subdivisions were created," complained county commissioner Dorothy Flowers in 1976, who estimated the price to fix the roads at $30 million, "there were no standards for the streets. In my opinion, a lot of developers pushed out the palmettos and put a little blacktop down and called it a street. Now they've got to be fixed," and "I get more complaints about bad roads than anything else."[84]

Eventually, in 1977, the county and the company came to an agreement: The county would accept roads in populated areas of Port Charlotte if—and only if—GDC built or resurfaced the roads to county specifications. That was a "step forward," said one commissioner, who admitted, though, that no matter what GDC did in the future, the county's road-maintenance program was a fiscally unmanageable mess.[85] The county had too many roads and not enough taxpayers to pay for them. There were hundreds and hundreds of roads, indeed vast networks of roads, with street signs and stop signs and ditches and culverts, but no houses or people, as if the military had triggered a nuclear device there, collected its data, then left. In November 1996, *Sarasota Magazine* columnist Bob Plunkett drew the ire of North Port officials when he called the city's "desolate, undeveloped outback" the "creepiest place in Florida," then added that it was the "perfect place to dump a body." Sensitive to bad press and seemingly oblivious to the fact that a suspected serial

To keep down costs, Florida's land giant developers built roads to their homesites using a crack-prone mix of Portland cement and local soils that residents called "grassphalt." Courtesy of Kevin Borland/Flickr.

killer had actually been dumping bodies there, North Port officials demanded (and received) an apology.[86]

Yet, whatever the case, for the past fifty years or so, the land giant communities' most defining characteristic has been their roads—their cracked, bumpy, trash-strewn, grass-filled, eerily deserted roads—to nowhere. "From the air [the roads here look] like a crossword puzzle with no clues," wrote a *News-Press* reporter in 1975. "They don't appear to go anywhere, yet . . . they seem to be everywhere. Miles and miles of roads, a maze of pavement and stop signs and cleared lots without a house in sight. They're a Florida phenomenon symbolizing the dramatic growth of an area and the population explosion to come."[87] In some places, like Port St. Lucie, the explosion did come, but for Citrus Springs or Port LaBelle or Rotonda, among others, roads are plentiful but residents are few and far between.

The southwestern part of Palm Bay, Florida, a city of 110,000 in Brevard County where GDC's Port Malabar development was located, is a case in point. Dubbed "The Compound" by locals, it's a completely barren, 5,000-lot

The goal of 1950s-era land developers was to subdivide their properties into as many 80′ × 125′ residential lots as possible. As a result, some sixty to seventy years later, portions of many land giant communities, including, most notably, southwestern Palm Bay and eastern North Port (pictured above), are still eerily empty. Courtesy of Google Earth.

wasteland with some 200 miles of roads, all crumbling, which were given to Palm Bay by default when GDC went bankrupt in 1991. The company, which grew out of the 1954 partnership between the Mackle brothers and Florida Canada, had survived the break with the Mackles and the recessions of the 1970s by shifting its focus from lot sales to homebuilding. During the first half of 1979, for example, while Gulf American's successor GAC Properties was reorganizing itself through Chapter 11 bankruptcy proceedings and the cash-strapped Deltona Corporation paid down its debt by selling the centerpiece of its Marco Island development, the Marco Island Hotel, GDC sold more homes (1,280) than at any point in its twenty-five-year history. In fact, the company created by the Mackles would outlast the Mackles. In February 1986, an aging Frank Jr. and son Frank III resigned from the Deltona board after years of losses—some $55.8 million between 1982 and 1984—and a controlling stake in the company was purchased by Minnesota Power and

Light, which had bought Deltona, tellingly enough, not for its building expertise or its land, but for its utilities.[88]

Meanwhile, GDC had big years in 1986 and 1987, posting profits of $22.6 million and $23.1 million, respectively, while building some 3,000 homes. With 5,000 employees and more than sixty sales offices worldwide, the company owned 267,500 acres in Florida in the "Port" communities of Port Charlotte, North Port, Port St. Lucie, Port Malabar, Port LaBelle, and Port St. John, plus Vero Beach Highlands, Sebastian Highlands, Julington Creek, and Silver Springs Shores. With assets of $1 billion, its businesses included home and homesite sales; the management of commercial properties such as strip malls and shopping centers; the financing of mortgages; the delivery of utilities; and the sale and administration of time-shares, of which GDC was the largest provider in the United States. Yet, no matter how much it diversified, GDC was a swamp peddler, an evil empire whose greedy, grasping, and unethical sales practices earned it the nickname "Generally Defective Communities."

For whereas other companies, such as Deltona and Avatar, had given up the practice of selling lots through installment plans, GDC upped the ante. It offered incredibly long plans spanning ten and a half and even twelve years, then used the contracts to upsell customers on homes that were overpriced. The process was simple: First, the GDC sales network would push lots the old-fashioned way, through bird dogs, cold calls, and sales dinners, then once customers signed they would be bombarded with personalized, in-home sales pitches urging them to "act now" because homes were appreciating so quickly, salesmen said, that customers were losing out on equity by waiting a decade to build. But GDC promised to help. It would bring customers to Florida in a heavily subsidized "Southern Ho!" community inspection program, where, after being wined and dined and spending three full days with a salesman, they'd trade their unimproved lots for a lot with a home.

Buyers would even get a rebate on the home's price and could list the home for rent through a GDC subsidiary called Florida Home Finders until they were ready to move. Unfortunately, GDC's homes were wildly overpriced, having been appraised by GDC-paid appraisers who set values at what other GDC homes had sold for, on average 20 to 40 percent more than they were worth. Then, GDC's mortgage subsidiary, the GDV Financial Corp., handled the mortgages, which it bundled and sold to financial institutions throughout the United States. In other words, fraud begat fraud begat fraud, with buyers scammed at the front end with faulty appraisals, then a knowing mortgage company accepting the appraisals and giving loans for them, then selling the loans out the back. The scam almost unraveled in 1984, comically so, when a California couple won a GDC home on the popular TV show "Dream House," which the company had valued at $85,000 and therefore the IRS required $20,000 in taxes for. The couple didn't have it, but when they went to a bank for a loan, they were shocked to discover that the home's real appraisal price was $40,000. (Faced with a public relations fiasco, GDC paid the couple $70,000 to be quiet.)[89]

There were plenty of other warnings and other settlements, but average buyers didn't know that. The average buyer was an immigrant or a retired worker from the north, or a foreign investor who hadn't been told that since the mid-1980s both the Federal National Mortgage Association (Fannie Mae) and the Federal Home Loan Mortgage Corporation (Freddie Mac), the nation's largest lenders, had stopped buying GDC mortgages. They were inflated, the agencies said, with Fannie Mae taking the extraordinary step of requiring GDC to buy back any mortgages it previously purchased that had gone into default. "When a company turns off a huge institution like Fannie Mae obviously [it's] done something wrong," explained Thomas Powers, a real estate expert at the Goodkin Research Corp. of Fort Lauderdale. There's "a cloud that's hanging over their operation."[90]

Yet, back in Florida, as customers toured communities on carefully scripted "Southern Ho!" programs, GDC's home-selling skies were remarkably clear. In fact, salesmen painted a rosy picture of property values—meaning they lied—while company bosses instructed workers to keep customers, known as SoHos, in the dark. GDC "taught us to make certain that there were no newspapers at motels," insisted Dawn Hamm, a Port Charlotte real estate agent who worked for the company in the 1980s, "so customers couldn't . . . [see] prices of other homes being sold or built." GDC also paid motel workers, she said, "not to put through to [customers] any calls from . . . local realtors or builders," and to keep customers busy, viewing lots, touring homes, and signing contracts, hopefully, until the moment they left. Hamm, who testified in court in 1990 and who also appeared on the investigative program *60 Minutes*, said, "It [was] clear to me that GDC's exchange program [was] a carefully devised scheme" to trick people "into buying GDC homes. . . . It [was] one big lie," she explained, and that's "why I quit."[91]

Eventually, news of GDC's pilfering found its way into the press. In February 1987, for example, the *Miami Herald* ran an expose followed by a mocking editorial by famed Florida humorist Carl Hiaasen after it learned that the state's Division of Land Sales had received no fewer than 400 complaints concerning GDC and its appraisals in three years. "Well, PARDON US FOR LIVING, OK?" joked Hiaasen, because "I'm sure there are excellent reasons for these minor discrepancies. So much can happen to a new house in eight months or a year—the paint can fade, the dog can mess up the carpets, the sprinklers can turn the sidewalks orange. Fifty bucks here and there, and before you know it, you've got $24,000 worth of serious depreciation. More to the point, why would anyone want to sell their lovely GDC home anyway? The whole idea is to move to Florida and spend eternity in paradise, assuming the roads eventually get paved."[92]

In April 1988, a federal grand jury in Miami began hearing evidence that GDC officials had not only deceived customers but had defrauded banks and the mortgage institutions Fannie Mae and Freddie Mac. Then, the infamous Mark Binstein got involved, as did more than 100 individual owners whose lawyers filed suit, plus two different class-action groups in Miami and a large (though non-class-action) group in New Jersey representing, all told, some 90,000 plaintiffs.[93] For its part, GDC claimed it had done nothing wrong. Sure, its homes were pricier, officials said, but the company had wholesale marketing and infrastructure costs, including roadbuilding, that local companies did not. "The fact that GDC homes may be more expensive than other homebuilders' houses is not a secret," claimed company chairman David Brown in 1990. "These other builders—we call them free riders—have not had to bear the enormous expense of building communities."[94]

That made sense, on the surface, but then why the bogus appraisals? They became so bad—with GDC houses going for $55 a square foot when comparable houses went for $30 to $34 a square foot—that in mid-1985 all four of its appraisers quit. "It was getting so far out of line that it just wasn't reasonable anymore," said one appraiser, who added they couldn't work for the company "and continue to be ethical" at the same time.[95] Nevertheless, GDC continued to sell conspicuously inflated homes, bringing in more revenue through the first three-quarters of 1989 than during any other year in its history, an astronomical $511.6 million. Its profits were down because of timeshare purchases, but its homesite sales ($264.7 million) and its housing revenues ($91.3 million) were up. In fact, and in spite of its legal problems, in 1988 GDC attempted to buy both the Deltona Corporation and the Land Resources Corporation, the owner of Lehigh Acres, which would have created a land sales juggernaut of some twenty communities 400,000 acres in size.[96]

However, in the fall of 1989, things started to go south for the company when out of the blue, GDC's South Korea sales chief, Ray Pionikoski, was arrested. It seems Pionikoski had merely done what GDC salesmen in Korea had always done: sell overpriced lots to GIs. But, when he and his swamp peddlers targeted Koreans, taking in some $5 million in payments for what journalists in Seoul dubbed "reed land," the Korean National Police intervened. It took nearly a full refund, plus the hurried interventions of Florida Senators Bob Graham and Connie Mack III (the son of former Gulf American official Connie Mack Jr.) as well as U.S. Representative and Foreign Affairs Committee Chair Dante Fascell, to get him out of jail.[97] The real bombshell, though, came in January 1990, when GDC reported to the Securities and Exchange Commission that the federal grand jury investigating its mortgage practices was about to indict the company and its officers for "possible violations of mail fraud, wire fraud, racketeering and related statutes in connection with certain aspects of the company's housing product pricing and disclosure practices."[98]

As soon as the news broke, GDC's stock price dropped precipitously, from over $9 a share in January to a mere $2.88 a share in March 1990. In all, the company lost $64 million in value in four months. Meanwhile, its long- and short-term bonds dropped almost in half, while Standard and Poor's lowered the company's debt rating from B to CCC, which meant it considered GDC as "not likely to have the capacity to meet its financial commitments."[99] Then, on 16 March, federal authorities in Miami announced a sixteen-count indictment charging GDC with a host of fraudulent practices that they claimed had earned the company some $850 million on more than 10,000 overappraised homes in Florida between 1983 and 1989. As part of a plea agreement, the government allowed GDC to plead guilty to a single fraud conspiracy charge while firing its executives and agreeing to refund $160 million to homeowners through a tedious reappraisal process run through the courts.

In April 1990, however, GDC declared Chapter 11 bankruptcy, which wiped out its restitution fund and forestalled its lawsuits, while leaving no fewer than ten local governments it owed taxes and improvements to in the dust. Palm Bay, for example, had 583 miles of roads to fix at a cost of $73 million. Likewise, Port St. Lucie had 230 miles to fix, but GDC owed the city another 163 miles plus culvert and swale work worth $96 million. Charlotte's tab was $30 million, later reduced to $8 million, which included roads and canal and seawall work that GDC had neglected.[100] "If the company goes under the city is in a crisis situation," said Palm Bay Mayor Frank Filiberto. "[We just don't] have the infrastructure to take care of those problems. . . . [It's as if GDC came in] and said 'We'll build you 900 miles of roads,' but didn't say the taxpayers would have to pick up the tab in 20 years."[101]

Unfortunately, like GDC's homebuyers, cities and counties were unsecured creditors, so as the company reorganized in 1991 under the name Atlantic Gulf Communities, banks and mortgage houses received payments in cash while unsecured creditors received stock. "I can't think of anything less that my clients would want than stock in GDC," said Doug Lyons, a Miami attorney who represented some 76,000 lot and home purchasers in lawsuits. "That's like adding insult to injury."[102] Yet cities and counties did have an ace in the hole: They could use GDC's bankruptcy to file condemnation suits to force it to sell its utilities, which Charlotte, Palm Bay, North Port, and Port St. Lucie each did between 1990 and 1992. As Tim Pozza, a Charlotte County utilities director, explained: "We've never been successful in dictating" to GDC "where [it should] put water and sewer lines." As a result, the company's been "determining [our] growth, not county officials. So the only way to resolve the problem is to buy the utility yourself."[103]

Likewise, to recoup their infrastructure money, counties and cities could place liens on GDC properties to stop it from building homes, but Palm Bay, Port St. Lucie, Port Charlotte, and other GDC communities were so building-dependent, so steeped in the construction trades and dependent on property taxes that they couldn't afford it. For as it stood, GDC had 353 employees in Charlotte County alone, where it had built 200 to 400 homes a year for the past thirty-five years and owed $2.4 million in taxes.[104] It owed Port St. Lucie $1.7 million in taxes, Palm Bay $1.3 million in taxes, all the way down to poverty-stricken Glades County, where portions of Port LaBelle were located and where GDC's unpaid $260,000 tax bill left a huge hole in the budget. Things were so tight in Glades that in 1990, the *Palm Beach Post* reported that Sheriff Chuck Schramm could no longer afford guards for his jail's graveyard shift, so each night, he left twenty-one prisoners, including an accused rapist and an attempted murder suspect, locked in their cells but otherwise unguarded.[105] Things weren't as bad in the other communities, but in North Port, where GDC owned perhaps half of the city's land mass, property taxes jumped 21 percent in 1991 and fire and rescue fees increased by 57 percent.[106]

Therefore, the only solution for North Port and the other cities and counties was to assist GDC with its reorganization process—to get it back to work, so to speak—by settling, not for stock in Atlantic Gulf, of course, but for the only real asset that it had: land. Thus, in a series of land-for-improvements settlements, Port St. Lucie took 1,600 acres; Palm Bay, 781 acres; Charlotte, 472 acres; and so forth. Palm Bay received the GDC Welcome Center on U.S. Route 1 and Port Malabar Boulevard where, locally at least, it all began. Meanwhile, homeowners received pennies on the dollar, specifically in nonguaranteed bonds maturing in 1996 and 1998 and Atlantic Gulf stock. But, in 1993, homeowners did get some good news: Former GDC president Robert Ehrling, chairman David Brown, and marketing executives Tore DeBella and Richard Reizen went to prison. Yet, even then, not everyone was excited about it. "Putting them behind bars doesn't do us any good," explained Palm Bay Mayor Bob Devecki. Instead, "they should have [been forced to work on] the road crew" to "fill all the potholes . . . they've left us."[107]

CHAPTER SEVEN

# Foreclosure Tours R' Us

> [They're] stucco ghettos. I live in one, so I can say that. Too many houses, not enough water, the economy's terrible, no tourists. This is the capital of the low-wage jobs, and when things go bad people have no safety net. . . . This is the epicenter of everything that's bad in America.
>
> —DOUG BENNETT, *St. Petersburg Times*

In the mid-1980s, former Tampa-mayor-turned-mall-builder Dick Greco visited Port Charlotte with friend and boss Edward J. DeBartolo Sr. to inspect the site of the future "Port Charlotte Town Center," a 100-store, 1.3 million-square-foot shopping facility on U.S. 41 in Murdock. According to Greco, the billionaire DeBartolo, who owned both the Pittsburgh Penguins hockey franchise and at least fifty shopping malls nationwide, including twenty in Florida, liked what he saw. Charlotte County was growing, booming even, and had added over 60,000 new residents since 1970. But, "as we toured the place," remembered Greco, "we said to ourselves, 'Man, these people are old! I mean, really old, like octogenarians.' That's when I told Mr. DeBartolo that growing up in Tampa we used to say people in St. Petersburg were old—you know, we'd joke and call the city 'God's Waiting Room' and things like that—and that's when he turned to me and said, 'Well, if the folks in St. Petersburg are old, then these are their parents!' We just about died."[1]

Greco, who still lived in Tampa, had been mayor of a city where the median age was thirty-two. By contrast, the median age in Charlotte County, just 100 miles to the south, was fifty-seven, which made it the oldest county not just in the state but the nation. "Silver and Old," read one 1987 headline in the *St. Petersburg Times*, "Nation's Oldest County Is Heaven for Retirees, Dullsville for the Young." Journalist Cynthia Mayer wrote, "It's dusk in Charlotte County, and Sheriff's Deputy Robert Brown looks out his windshield. Ahead of him is a handful of the town's teenagers, laughing in a tight circle. Behind him is an empty mall, and behind that, thousands of ranch houses, most occupied by senior citizens. Senior citizens who would like to get to sleep soon."[2] Mayer was in Port Charlotte, interviewing teens at an outdoor mall owned by the General Development Corporation (GDC), where they told her, repeatedly, that the community was so retiree-oriented, so focused on providing the elderly with recreational activities and services, that there was nothing for young people to do.

"After dark you've got two choices here," said one resident. "You can sit and watch the puddles dry, or you can sit and watch the cars rust." Indeed, in 1987 there were roughly 8,000 school kids in Charlotte County, but no Boys and Girls Clubs in Port Charlotte, no YMCA fitness centers or gymnasiums, and believe it or not, no video game arcades. However, when commissioners voted to implement a $12 annual tax on lot owners to buy land for parks to actually give kids something to do, an estimated 500 enraged retirees attended the next commission meeting, where they beat on the room's glass walls and tore a door from its hinges. Most of the protesters had fixed incomes, they yelled, and though property taxes were low in Charlotte, and Florida had no state income tax, they refused to pay for services that they themselves didn't use.

"You and I both know that the majority of park users are from the younger group," exclaimed local resident Ceil Gates. "We build the parks for them, and they vandalize them. We [senior citizens] can no longer afford this or any other increase in our taxes."[3] Like Gates, budget-conscious Floridians expressed similar sentiments statewide, but those in land giant communities like Port Charlotte, where many residents were fixed-income retirees, were notoriously anti-tax. "We're living payday to payday as far as county operation and planning are concerned," said Charlotte County manager John Printon in 1982. "Long range planning is very difficult. Nobody has any interest it. The people question why they should commit themselves to something fifteen or twenty years down the road."[4]

In Hernando County, where the Deltona community of Spring Hill was located, some 54 percent of the county's 67,000 registered voters were age 60 or older, and in November 1990 they voted "no" on a tax referendum that would have raised money for parks as well as a badly needed social services agency for children. As they had in Charlotte, they packed commission hearings, where they told officials, "I was a Depression baby, and it was damn tough!" and "We just can't continue to be taxed to support projects that we have nothing to do with," and shouted down a sixteen-year-old high school student who had temerity to speak. "It got pretty nasty," said county administrator Chuck Hetrick. "I don't think it's universal, but a lot of your elderly people feel they paid their dues to the young people and want to be left alone."[5]

In 1988, for example, residents of the 2,400-unit modular-home community of Brookridge north of Spring Hill voted to restrict the community to adults age 55 and older. Brookridge, which had been open since the early 1970s, had just thirty children and teens living there, but its lot owners voted 1,551 to 949 to keep any more of them out (they also voted to bar anyone age 18 or younger from even entering the community and to give any woman who gave birth to a baby while living there a year to leave).[6] In the city of Tamarac, Florida, in Broward County, first developed in 1962 as an adults-only community by construction magnate and future Seattle Seahawks owner Ken Behring, residents of newer, all-ages neighborhoods on the city's western edge reportedly bussed their kids to eleven different county schools, outside of Tamarac, in six years, including a junior high school ten miles away in Pompano. "If I were to do it over," said a Tamarac mother, "I wouldn't come here again. It doesn't have schools, it doesn't have parks, [and] it doesn't have government participation by so-called 'young people.'"[7]

But that was the rub: Large numbers of retirees concentrated in planned communities were politically active and had clout. For one, people age 55 and over registered to vote in greater numbers than people age 54 and younger did—78 percent versus 65 percent in the mid-1980s. Then, those age 55 and over voted at higher rates, 70 percent to 56 percent, which meant their political influence was far greater than their size.[8] In 1992, for example, residents age 55 and over in Palm Beach County represented 53.6 percent of registered voters but just 34 percent of the population.[9] Likewise, in 1989, residents age 65 and over in Broward, Charlotte, Hernando, Indian River, Lake, and Marion counties were a collective 26 percent of the population but a whopping 41 percent of the vote.[10] "Vociferous, ever-present, and often demanding," wrote the *Tampa Tribune* in 1987, "people in large retirement communities often show up en masse at county commission meetings and cow their leaders," which is what happened in Charlotte with the park tax, when some 1,500 angry pensioners, guarded by no fewer than ten sheriff's deputies and three bailiffs, attended a follow-up meeting at the county auditorium and forced the commission to reconsider.[11]

However, politically, older Floridians were a mixed bag. In local referendums, for instance, they generally opposed school bond issues and park taxes (especially if they had lower incomes) but supported increased Social Security and healthcare benefits at the national level. And, depending on where they were from and how they voted in their preretirement years, they brought their party affiliations with them. In Southwest Florida, for example, where Interstate 75 connected Collier, Lee, Charlotte, and Sarasota Counties with the generally conservative Midwest, senior voters were and continue to be Republican, wildly so, while major planned communities at North Port, Port

Charlotte, Punta Gorda Isles, Cape Coral, and Marco Island, plus wealthy enclaves at Naples and Sarasota, have made it perhaps the most GOP-friendly region in the nation.

"The growth in [the Republican Party here] is from migration, not from new converts," explained Norman Scherman, a GOP official in Sarasota County in 1986. "About 70 percent came here from Ohio, Illinois, Michigan, places like that, where they were always Republicans."[12] Thus, as Southwest Florida grew and its planned communities grew, the Republican Party surged to the fore. It surged, too, in Pinellas County, the retiree-heavy region surrounding St. Petersburg, which by 1988 had more registered Republicans than any other county in the state, and on the Nature Coast, where land giant communities at Spring Hill, Citrus Springs, and Sugarmill Woods, among others, swung party affiliations and voter demographics to the Republicans.[13] The numbers didn't lie. In 1956, for example, registered Democrats outnumbered registered Republicans in sparsely populated Lee County 17,531 to 2,277, but some thirty years later, in large part because of conservative transplants to Cape Coral, which by that point was Lee's largest city, Republicans led Democrats 74,226 to 61,327.

And similarly, neighboring Charlotte County went from 3,336 Democrats versus 226 Republicans in 1956—the year Port Charlotte opened—to a 1988 total of 29,817 Republicans versus 22,982 Democrats.[14] In fact, Charlotte's first-ever Republican officials came from Port Charlotte, where in the late 1950s, a retired Republican auditor and salesman from Crawford County, Pennsylvania, named Donald Fisher, whom locals called "Mr. GOP," began a one-man organizational drive to, as he put it, "see that Republicans outregister Democrats in Port Charlotte." To that end, Fisher went from house to house, engaging every Republican resident he could find, then urging them to register after they lived in the county for one calendar year. He then formed the Port Charlotte Republican Club, whose members called, visited, or drove to the polls nearly every one of the community's 560 Republican voters, who in 1960 elected Republicans to both the state legislature and the county school board.[15]

Florida's Republican growth trend, started, as it were, by retired "pioneers" like Donald Fisher in Port Charlotte, wasn't just a Gulf Coast phenomenon. On the contrary, by the 1980s, hundreds of thousands of senior Republicans had settled in a "horseshoe-shaped corridor" from Collier County in the south, up the coast through St. Petersburg, over to Orlando then down the Atlantic coast through Brevard, Indian River, and Palm Beach counties, until they were met, finally, and politically diluted by hordes of liberal Northeastern pensioners in towering condominium complexes in Broward and Dade. (In the same way that I-75 had connected Southwest Florida to conservative parts of the Midwest, I-95 had connected Democrats from New York and New Jersey with Miami.) In rough geographic terms, Florida's Republican horseshoe contained the state's main land giant communities, with those on the west coast (Cape Coral, Marco Island, Port Charlotte) being more reliably "red" than those on the east coast (Deltona, Palm Bay, Port St. Lucie).

The reason was that, while both sets of communities had large numbers of GOP-leaning retirees, Florida's west coast communities were older, more dependent on fixed incomes, and less dependent on nine-to-five jobs while developments on the east coast were bedroom communities for the young. In 1985, for example, the average Port St. Lucie resident was barely thirty-four years old. These younger residents sometimes worked in the community but more often than not commuted to jobs in Stuart, West Palm Beach, or Fort Pierce. "Port St. Lucie used to be known as the home of the newlywed and [the home of] the nearly dead," said city councilman Pat Kelly, but "as the city has grown, there has been a shift in the population. There are more and more working people in the community."[16] Likewise, one 1990 study found that in Deltona, which the Mackles had opened as a planned retirement community in 1962, some 5,000 cars headed each morning through the I-4 interchange to Orlando. Deltona had one exit in those days, so traffic jams at the on ramp were sometimes a mile long.[17]

No doubt, what drew commuters was the price, because lots were plentiful here, almost superfluous, and as a result, homes were cheaper in Deltona than in the first-ring suburbs of Orlando, so workers made the drive. "Deltona still has the best house buy that I'm aware of" in the Orlando area, said a Century 21 rep in 1990. "It's difficult [here] . . . to find an 1,800-square-foot, 3-bedroom, 2-bath brick home with nice landscaping priced in the $80s, but you can in Deltona."[18] In Palm Bay, where workers commuted to the Cape Canaveral area, homes were even cheaper, sometimes in the $50,000 to $60,000 range, while those elsewhere in Brevard averaged a far more expensive $125,000. Nevertheless, low home prices notwithstanding, the shift in Palm Bay, Deltona, and Port St. Lucie from retirement community to bedroom community was an arduous one. In all three places, the lot-sales-focused, 1960s-era planning was poor: there were no downtowns or youth centers, no big employers or nightlife, while in Port St. Lucie, three of the city's five sitting councilmen, including its mayor and vice mayor, were retired.

"I think they're totally out of touch with the needs of the community," said a local youth activist. "The majority of people in Port St. Lucie are under forty, but the city council is owned by [older residents]" and "is scared of their votes."[19] The city had extremely stiff noise ordinances, for one, prohibiting loud music from emanating from bars within the city's thin commercial district between the hours of 11 P.M. and 7 A.M., and once made national news when it issued a court summons to the manager of Rooster's Seagrill for allowing a karaoke party to sing "The Hokey Pokey" too loudly. City police then cited the Club Med Village Hotel of Sandpiper, one of the few large businesses within Port St. Lucie, for welcoming a bus full of guests after 11 P.M. with what an officer reported as "clapping, chanting, and occasional music," and issued not one but four citations to a thirty-two-year-old resident who held an afternoon party with a rock and roll band in his yard. For ignoring the city's noise ordinance, which among other things specifically prohibited squawking parrots, a judge sentenced the resident to fifty hours of community service and six months of probation.[20]

Then, in the mid-1990s, the city council unlawfully built a senior center and a senior bus depot on a city park site zoned "open-space recreational," while, in the process, ignoring its own survey that found 50 percent of respondents preferred a youth center while just 36 percent had favored a center for seniors.[21] However, Port St. Lucie did have a Major League Baseball spring training facility, including a stadium and clubhouse, where the New York Mets played, but that had been paid for not by the City of Port St. Lucie—which was so anti-youth that the city once fined a family $200 a day for having a skateboard ramp *in their yard*—but by a private developer and a 2 percent hotel tax sanctioned by the county.[22] The Mets' stadium provided badly needed entertainment to residents of all ages, youth included, but New York sportswriters who spent six and a half weeks in the city each year found it boring, sometimes referring to it in print as "Port St. Lousy."

The famously nasty *New York Post* reporter Andrea Peyser went even further, calling it "an ugly town on a numbing stretch of faceless highway lined with shopping centers and honky-tonk bars serving fresh seafood, cold beer and briefly clad babes," adding that Port St. Lucie was "a pathetic waste of map space."[23] Yet, what did Peyser expect? The Mets had moved to Port St. Lucie from St. Petersburg, which at the time was itself a retirement mecca, where one writer quipped the only "movers and shakers" there were paramedics.[24] And besides, Port St. Lucie was platted that way, as a single-family residential community with no downtown that had grown far too big. In 1980, for example, the city had just 14,690 residents but no high school or police force—it actually rented deputies from the county—but ten years later it had ballooned to nearly 56,000 people, an astonishing 280 percent increase and the equivalent of eleven new residents a day.

Thus, with no central core and limited condos and high rises, Port St. Lucie sprawled and sprawled and sprawled; *New York Times* travel writer Sam Roberts called it "the quintessence of suburban sprawl, but with no big city to sprawl from."[25] That meant that the onus of city services rested with homeowners because Port St. Lucie's commercial base was so pitifully small, while the city itself lacked an identity. It had no cultural district, no center, and an older voting population so preternaturally averse to taxes that it would never build one. "A lot of people say, 'Well, where's your downtown?'" said former mayor and city councilman William McChesney in 1991, to which "I always say 'I hope and pray we never have a downtown.' If you look at a lot of cities, the downtown is the first thing to go wrong."[26]

But, without a nucleus, there was "no there there," as novelist Gertrude Stein once wrote, no sense of place. It wasn't until 1992, some thirty years after the city's founding when its population had hit the 60,000 mark, that a hospital in Port St. Lucie added a maternity ward. Finally someone could say, "I'm from here"—a baby boy named Matthew weighing seven pounds, ten ounces.[27] Yet, for everyone else, Port St. Lucie was a transitory place. It lacked a cemetery until 2004, for instance, because families of deceased residents sent loved ones "home" to be buried, up north, or to surrounding communities, while the city itself had few big nonresidential properties to begin with and therefore nowhere to put it. "I remember asking when we first moved here," said local resident Mary Lemon, "'where's the cemetery?' and I thought it was strange that a city this size didn't have one."[28]

Things were similarly impermanent in Spring Hill, the Deltona development in Hernando County, where the community's only county-designated historical landmark was a waterfall and fountain built by the Mackles in 1968. However, by the mid-1990s, Spring Hill had at least 50,000 residents, including an increasing number of Tampa-bound commuting residents, a Kmart, a Target, two Walmarts, and three McDonald's, but just a single public library, an incredibly small, 3,000-square-foot facility housed in a former Deltona sales office.[29] While Spring Hill celebrated its past with an annual chicken-plucking festival, a tradition that began as a Deltona public relations stunt in 1970, the 150-year-old city of Brooksville, the "Old South" seat of Hernando County just twenty minutes away from Spring Hill, held "Cracker Days" and "Heritage Days" and reenacted a raid from the Civil War. "The people who live out in Spring Hill, some of these people practically never,

ever get to Brooksville," said a longtime local resident in 1990. "They have their own culture, and their culture is that which they brought down with them. They don't really know or understand or attempt to understand what they have moved to. In effect, they have never become part of the county."[30]

In the mid-1990s, most Spring Hill residents would have agreed with that. "We're a transplanted population that's unwilling to cut the placenta," admitted retiree and former Spring Hill civic association president Nick Morana, and "we don't know each other. We come from various neighborhoods and various states. We're cliquish. But the salvation of the whole situation, I think, are the young people who are moving in with their children. And if they stay here, I think things will be better."[31] Eventually things did get better, community-wise, as thousands of young families who moved to Spring Hill stayed. Some had followed the grandparents to Florida and had taken jobs in the service sector or in healthcare or construction, while others had escaped high home prices in Tampa for life in the far suburbs; the savings were that great. In 1997, for example, the average home price in Hillsborough County was $112,414, while in Hernando, it was a shockingly low $66,851.[32]

Although by 2000, the median age there was still a very old 49.5, the number of residents age 19 and younger grew 31.4 percent, from 20,591 in 1990 to 27,050 in 2000. Likewise, residents in the 35–44 age range, who presumably were their parents, grew an impressive 42.7 percent, from 10,967 to 15,653. "We've been seeing an influx of younger people" and "people in their forties that have children in the high school ages," said one realtor, who added she no longer marketed Spring Hill as a place geared for people fifty-five and older.[33] Indeed, the "new" Spring Hill had two youth centers and a skate park, while in 2000 commissioners approved a five-year, $4.7 million expansion program for the county's park system, which added, among other things, a swimming pool, numerous baseball fields and a soccer field, some basketball courts, and a tennis and racquetball complex.

Yet, like Port St. Lucie, Spring Hill's youth movement wasn't without its difficulties. In 1995, a local skating rink operator attempted to add an alcohol-free teen club to his business, called "Sponges," by signing a licensing agreement with popular Tampa shock-jock "Bubba the Love Sponge" Clem, but opposition by residents, who signed a "Keep Bubba Out!" petition and pointedly claimed that love sponges were "inappropriate" role models, forced the operator to close.[34] In 1999, the Love Sponge returned to Spring Hill with an over-21 dance club that he himself owned called "Planet Bubba," whose raucous crowds and thumping base disturbed residents next door. However, the sound levels were legal—inspectors would measure the noise at least a dozen times, to no avail—while county officials were the ones who had allowed Deltona to build homes next to a strip-mall business district in the first place. Nevertheless, older, conservative commissioners targeted the Love Sponge, whose popular radio segments included "Lesbian Tuesday" and "No Panties Thursday," and in 2000 tightened the county's "unlawful sound" ordinance by lowering decibel levels and making each offense a misdemeanor in order to shut down the Love Sponge. "It's just a bunch of old biddies pitching a fit," said a Planet Bubba patron. "They don't have anything better to do than to sit around and bitch."[35]

Yet dance clubs or no dance clubs, by the early 2000s, young people were moving to Spring Hill in droves. The reason was that in 2001, the state's Florida Turnpike Enterprise finished the forty-two-mile-long "Suncoast Parkway" connecting Spring Hill with Tampa, which, for a toll, allowed commuters to zip south to the city's Veterans Expressway, then from there past Tampa International Airport to I-275 and beyond. End-to-end, the trip took fifty minutes, and soon thousands of Spring Hill residents were making the commute, an estimated 9,600 cars per day.[36] "My friends in Tampa are like, 'Oh my gosh, you live in the country,'" said a twenty-seven-year-old Spring Hill resident. "But this is not the country anymore. This is *not* the country."[37]

Spring Hill was, by that point, a bedroom community for Tampa, where cheap home prices, low mortgage rates, and a willingness among lenders to give mortgages to low-income borrowers allowed people who had once only dreamed of owning a home a chance to actually do it. In fact, as of 2002, no other county in the state had a higher percentage of home ownership than Hernando, an amazing 86.5 percent, and many of those were in Spring Hill. "How wonderful to be a daycare worker or a clerk in a store and to be able to buy a house," said a gushing Hernando real estate agent. "That's only going to make this area better. People have more pride [in their community] when they own their own homes."[38]

In Southwest Florida, in Golden Gate and Lehigh Acres in particular, the trends were similar: By 2000, these former retirement communities had grown significantly younger, to a median age of 29.6 and 38, respectively, becoming bedroom communities for service workers and construction workers who commuted to jobs on the coast. "[They're] the pool cleaner" and "the lawn maintenance guy," explained Bonita Springs Mayor Jay Arend in 2002, whose increasingly upscale city near Naples had few cheap places to live. "It's generally the gated communities, the expensive homes, that employ these people," and "they need housing."[39] With access to I-75, and upwards of 150,000 lots available, Lehigh Acres and Golden Gate drew thousands of low-income workers, including, most notably, Hispanic workers from Puerto Rico and Florida's east coast, whose numbers in Lehigh Acres rose from 585 in 1990 to 4,466 in 2000, an increase of 663 percent.

"We've sold a lot of homes to Hispanics," said Pat Logue, president of sales and marketing at First Home Builders of Fort Myers, which in 1998 purchased 714 lots in Lehigh Acres to build homes for low-income buyers.[40] "We can put [people in houses] for under $80,000. The turnkey price is $79,900 for a 1,300 square foot home with 3 bedrooms, 2 baths, and a 2-car garage."[41] Logue could do this because lots in Lehigh Acres were cheap and because he specialized in directing borrowers to mortgages insured by the Federal Housing Administration (FHA), a government agency that assisted first-time buyers in securing a loan—and, in particular, those with low-to-moderate incomes and not enough money for down payments. Whereas banks and credit unions typically required mortgage insurance for loans with down payments of under 20 percent, while in turn mandating down payments of at least 5 percent, most private insurers refused to guarantee loans from borrowers who were a risk.

That's where the FHA came in. It offered insurance to borrowers who could only afford down payments of between 3 and 5 percent, depending on the size of the loan, and would include the price of insurance and some closing costs in the mortgage. Of course, FHA insurance was more expensive than private insurance, the paperwork was copious, and there were limits to what people could borrow. In 1998, the maximum was $112,100 in Lee County, though it varied from area to area. "Anyone who qualifies for a conventional mortgage probably would prefer to go that route," explained Layne Morrill, president of the National Association of Realtors. "It's cheaper, it's faster, and it's easier to close."[42] Yet, for everyone else, including the 80,000 apartment renters then living in Lee County, where basic one-bedroom units cost between $440 and $689 a month and three-bedroom units started at $600 a month, FHA loans were a godsend, allowing those who qualified to buy (rather than rent) a landscaped house with even the window blinds preinstalled for as little as $580 a month, including principal, interest, taxes, and insurance.

The trick, realized Logue, a forty-year-old Missouri transplant and former used-car salesman in Fort Myers, was to make the down payment as small as humanly possible, which would allow buyers living week-to-week and paycheck-to-paycheck to secure a loan. After all, most could afford the mortgage because, at less than $600 a month, it was the same as rent. But few had $4,000 or $5,000 lying around for a down payment. So Logue made things easier for them. He paid closing costs and fees and up to $3,000 in buyers' debts; allowed them to delay down payments until after construction was completed; and offered a "no payment" grace period for the first sixty days. Then Logue connected buyers with a little-known nonprofit called the Nehemiah Progressive Housing Development Corporation, a Sacramento-based agency founded in 1994 by a Baptist minister and real estate lawyer named Don Harris, whose goal, at first, was to support urban renewal by forming a charity that would assist African-American buyers in Sacramento with making down payments.

To do this, Harris convinced local builders to donate money directly to Nehemiah—say, 6 percent of a $100,000 mortgage. Then, Nehemiah would "gift" a 5 percent down payment to the buyer while keeping 1 percent for itself as overhead. The plan was a win-win, insisted Harris, as it allowed Nehemiah to conduct its redevelopment mission in and around Sacramento while buyers got houses and sellers sold homes. "To me this was a big opportunity," said Gloria Jackson, a sales clerk and single mother from Sacramento. "I didn't have a bunch of money for a down payment or closing costs and all that. Nehemiah made it possible. It's wonderful. I'm happy, happy, happy."[43] The problem, though, was that FHA guidelines prohibited sellers from making down payments for customers, the fear being that buyers would walk away from a home with nothing put into it or that sellers would add the down payment to the price, which would inflate mortgages.

And, though employers, unions, family members, bridal registries, and charitable organizations could each add to down payments legally, in June 1997, the FHA's parent agency, the Department of Housing and Urban Development (HUD), ruled that, though well meaning, the Nehemiah program was, in effect, a pass-through, a scheme by which sellers made payments for buyers "through a circuitous route." Therefore, HUD withdrew its support for the program. "I don't care how laudable it is," said a housing analyst with the New York investment bank Natixis Bleichroeder, "you're letting people into homes with no money at risk."[44] However, less than a week later, HUD reversed itself, but when it refused to give Nehemiah more than a six-month "demonstration" period, the organization filed suit while asking the powerful Congressman Bob Matsui of Sacramento, a ranking member of the House Ways and Means Committee, to intervene.[45]

Grudgingly, and without the support of HUD's Office of Single Family Insured Housing program staff, in June 1998 the agency issued a memorandum stating that the Nehemiah program "was not in conflict" with FHA guidelines and that it and "other programs [so] similarly structured" could proceed.[46] It was a fateful decision, one with broad, largely unforeseen implications, because Nehemiah had been sending representatives to real estate and mortgage conventions nationwide because its goal, said Ron Mellon, a marketing expert employed by the nonprofit, was "to get the word out to every builder and realtor in the country."[47] Clearly it did, because just days after the HUD memorandum, the Fort Myers *News-Press* published an article titled "Filling the Local Housing Gap: Program Would Allow Buyers Low Closing Costs," in which it explained the basics of the Nehemiah program and featured Pat Logue and First Home Builders in particular.

"It's a very well planned, well thought out, and workable situation," said Logue. "It'll be the most significant thing we've ever used to help someone who's stuck renting."[48] Yet, even without Nehemiah, First Home Builders had been booming. With 271 homes completed or permitted in 1997, it would average more than fifty permits a month through 1998. Propelled by high employment, low interest rates, a strong stock market, and what observers called the "most buoyant" consumer confidence index in decades, permits for single-family home construction jumped 8 percent in Lee County between 1997 and 1998, from 2,252 permits worth $307 million to 2,433 permits worth $365.6 million.[49] The increase came from low-income homes, the kind First Home Builders specialized in, in Cape Coral and Lehigh Acres in particular, where in June 1998 the company built seventy-two units in a month. Then, in August, it became the largest lot owner in Lehigh Acres, with an inventory of some 3,000 homesites, and that month sold its first Nehemiah program home, the first such home in the state.[50]

For a company and a housing market already on fire, the advent of the Nehemiah program was a burst of pure oxygen. By the end of 2002, for example, First Home was the largest single-family home builder in Lee County, having pulled permits for 933 homes worth $92.1 million, more than its two closest competitors, U.S. Home and Beazer Homes, combined. Then, in 2003, it pulled another 1,517 permits, followed by 2,457 permits in 2004, an astonishing one in six single-family home builds in what *U.S. Housing Markets*, a trade magazine, called the hottest market, per capita, in the country.[51] And, in 2005, *Builder Magazine* named it the fastest-growing private builder in the country, for the second-year in a row, and ranked First Home as the fifty-first largest homebuilder, by volume, in the United States—all from a ten-year-old company that built homes in Cape Coral and Lehigh Acres specifically, and that did half of its business through seller-backed down payment assistance programs such as Nehemiah.[52]

But then an odd thing happened. The boom in land sales that began in the 1990s stretched on into the 2000s; prices rose, drastically, and the 1,274-square-foot First Home houses that went for $79,900 in 1998 jumped to the $150,000 range by 2005, and those came *without a lot,* the prices of which had rocketed from $1,400 in 1998 to $35,000 in 2005. The boom wasn't just a Lee County phenomenon; people were buying and building all over Florida, in record numbers, and at prices that pushed low-income borrowers out of the market. Statewide, for example, the median sales price of both new and existing homes ballooned from $116,100 in late 2000 to $247,800 in 2005, a five-year, budget-busting increase of 107 percent.[53] Clearly, Florida was a huge part of the U.S. housing bubble, which stemmed from a caustic witch's brew of rising incomes, historically low interest rates, an underperforming stock market that bottomed out during the "dot-com crash" of 2000 and forced investors into real estate, and a willingness among banks, savings and loans, the FHA, and the government-sponsored Fannie Mae and Freddie Mac to lower their loan and underwriting requirements to put millions of new buyers into homes.

To that add booming populations across the Sunbelt, but especially in Florida, where the state added 834 new residents a day through the 1990s to finish the decade with 16 million inhabitants, an increase of over 3 million people, or 23.5 percent. Fully a third of newcomers went to Florida's already big mega-counties (Palm Beach, Orange, Broward, Hillsborough, and Miami-Dade), but upwards of a million others would settle outside of urban areas in counties in southern and central Florida with a surplus of buildable land. And no one had more buildable land than the 1950s-era planned communities founded by the land giants. Between 1990 and 2000, for example, Lehigh Acres grew an astonishing 145.6 percent, more than six times the state average, to 33,340 residents. Likewise, Spring Hill grew 122 percent to 69,079; North Port, 90.4 percent to 22,797; Port St. Lucie, 59 percent to 88,769, and so forth.

"With few exceptions," wrote the *Orlando Sentinel* in 2003, "the reason for the population boom in [these] communities is the same: they offer an abundance of cheap, vacant lots."[54] At first, prices were low—so low that retiring boomers and empty nesters began moving there from other parts of the state, selling more expensive homes in, for instance, St. Petersburg to live in Cape Coral. Yet retirees from the north had the same idea; so did Haitians and Hispanics from Miami, transplants from Puerto Rico, African Americans from New York, and workers from every northern state east of the Mississippi, plus "snowbirds" from Canada who wintered in Florida but returned each spring before their health benefits ran out. If it looked like a stampede it was, and even experienced builders used to the ebb and flow of Florida's construction-dependent economy struggled to keep up.

"Lehigh Acres is on fire," said Wes Brodersen, president of the EXIT Gulder Real Estate firm of Naples in 2004. "We've probably seen prices more than double in eighteen months. Golden Gate is on fire—same deal. People are waiting in line to buy homes."[55] There were, in fact, waiting lists for homes, and in some new developments, such as Aviano in Collier County, Hawthorne in Lee County, or The Grove in Broward County, buyers actually entered lotteries to win the right to purchase a home.[56] Such extreme demand, coupled with a construction boom in China and four hurricanes in 2004 that damaged some 700,000 housing units statewide, created a cement shortage and a drywall and roofing materials shortage that inflated prices and pushed home completion rates from three to four months in 1998 to upwards of twelve months in 2005. "There's no question that time frames have increased for everybody," said Robert Gazzoli, president of SeaGate Homes in Palm Coast. "Our volume has increased significantly this year, and it takes longer to build homes."[57]

It took money, too, as prices for construction materials soared across the board. According to the federal government's producer price index, between 2004 and 2005, ready-mix concrete jumped in price by 12 percent; clay tile, 6.8 percent; PVC piping, 12.6 percent; and paint, 6.1 percent. Builders saw huge increases in the cost of framing lumber, which jumped 47 percent in 2003 and 2004, then another 13.5 percent following Hurricane Katrina in 2005, so they hoarded the wood. They also hoarded plywood and drywall, which both doubled in price, and in some places even asked buyers to approve three different roofing tiles in case their first or second choices ran out. "It's just economics," said Mike Hoyt, vice president of development at the Lutgert Companies of Naples. "There has been so much building going on and the demand is so huge that you pay a premium for the products you need."[58]

High building costs were one thing; rampant "flipping" and widespread property speculation that inflated prices and drove up living expenses were another. Although estimates vary, in 2004 the National Association of Realtors claimed that 23 percent of homes sold in the United States were sold to investors. But in Florida the numbers were higher. According to the Hillsborough County Property Appraiser's Office, for example, between early 2004 and mid-2005 some 40 percent of homes sold in the county weren't registered with the appraiser as the owner's primary residence, a sign that many of the buyers were investors. (The others, presumably, were vacation-home buyers.) Then, in 2005, the mortgage analytics company LoanPerformance estimated that from January through April of that year, 57 percent of homes in the Naples metropolitan statistical area, which included Golden Gate, had been sold to investors or vacation-home buyers, the second-highest percentage in the nation. Fort Myers-Cape Coral was fourth at 48.6 percent, Punta Gorda-Port Charlotte fifth at 46.5 percent, and the Treasure Coast and Port St. Lucie thirteenth at 34.9 percent.[59]

Spurred by wildly increasing prices, which varied by area but in fairly representative Palm Beach County grew an astonishing 29.25 percent annually between May 2001 and May 2005, investors began buying and rehabbing old houses (i.e., traditional "flips") or placing deposits on new houses, preconstruction, to sell at a profit. The latter practice—that is, preconstruction purchases—were especially prevalent in Florida and were "a huge factor in all subdivision housing" in the state, said Brad Hunter, a consultant for the West Palm Beach industry information firm Metrostudy in 2004. "People are buying preconstruction and looking for a quick sale. The majority of them are ordinary people who know somebody who knows somebody who made $100,000 on a flip, and they decided to jump on the bandwagon."[60] Indeed, every Florida newspaper, it seems, had a story: the barber, the butcher, the insurance salesman-turned-property-investor now soon-to-be-millionaire tycoon.

"You've got to be willing to take a chance," explained a twenty-six-year-old Ritz-Carlton doorman to a reporter at the *Naples Daily News*, which checked property records and found that in 2005, the doorman had purchased a townhouse, preconstruction, in the newly built Stratford Place community near I-75 for $185,400 then had sold it for $270,000 just *fifteen days later*.[61] Likewise, the *Palm Beach Post* touted a band director who flipped houses and doubled his teacher's salary in six months; the *Bradenton Herald* reported on students of a house-flipping course at Manatee Community College; and the *St. Petersburg Times* ran a story about a Connecticut man who commuted back and forth to Spring Hill, where he bought and held houses for a month before selling them at a $10,000 markup. "The first one we thought, 'Maybe it's a fluke,'" he told a reporter. "The second one, 'Maybe we're lucky.' Now, it's rolling."[62]

Nevertheless, flipping was a nasty business. More than anything else, it drove up the price of what had been affordable housing. Experts insisted that for each $1,000 increase, some 25,000 Floridians could no longer afford a home.[63] In ritzy Collier County, where on average construction workers made $33,000 a year, the median sales price in April 2005 was $477,100, the highest in the state. So workers lived where they always lived, in rural suburbs such as Golden Gate, but then speculators hit the community with a frenzy of home flips. Between January 2004 and July 2005, for example, some 325 properties in the Golden Gate Estates section were bought and sold twice in three months. The median increase was $27,000, with a median market time of just forty days.[64]

"What's happening is single-family housing is pushing itself out of the capabilities of work-force housing," asserted local planning expert Wayne Daltry. "Land is getting scarcer and more expensive" but workers here "don't just . . . disappear off the face of the earth once their day's work is done. They continue on and need a place to go home to and we have to find affordable housing for them."[65] This was unlikely, as even mobile home lots in Collier frequently sold in the $50,000 range, and teachers, policemen, firemen, and nurses, among other professionals, could no longer find places to live. The average Collier teacher, for example, earned $41,558, and while "as a rule of thumb" people could afford houses that were roughly three times their salary, insisted Cormac Giblin, the county director for housing and grants, there wasn't a single house listed in Collier, in October 2005, for under $200,000.[66]

Collier was an extreme case, by one estimate the twenty-sixth least-affordable housing market in the nation, but ballooning prices in neighboring Lee County, the fifty-fifth least-affordable market in nation, were almost as bad. There, the median home price was $281,000. As a result, companies such as First Home Builders began leaving the low-end market and its FHA and Nehemiah buyers to sell more expensive homes, specifically to investors. And why not? Properties were going so fast in Lee, and at such high prices, that in 2005, the Florida mega-builder U.S. Home closed its Cape Coral showroom and stopped building individualized homes for buyers because it was quicker to build "speculation homes" to immediately sell on the market. "Speculation isn't really a good name for [those homes]," asserted Robert D. Knight Jr., vice president of a competing company called Paul Homes. "In this area there is no speculation. You build a home, and you will make money."[67]

However, First Home Builders went even further. The company built not only speculation homes to sell in numbers but, together with a local realty company, devised a shady investment scheme called the "First Home Lease Purchase Investment Opportunity," in which it sold investors leased homes, which were then bought from the investors by the people who leased them.

The process was this: First, the company would build a home in either Cape Coral or Lehigh Acres. Then, it would find a tenant, someone with poor credit that First Home promised to repair while the tenant paid rent for the period of one year, at the conclusion of which the tenant would buy the home. Meanwhile, the company sold the home to an investor, who paid First Home $5,000 to cover the down payment of a loan whose own monthly payments were covered by the tenant's rent. In exchange, the investors received a 14 percent return on the sale.[68]

It was a risky scheme, one dependent on high home prices and easy-to-get loans. But, as property boomed in Florida and people caught speculative fever, lending institutions did too. Banks, savings and loans, credit unions, mortgage companies, insurance companies, and brokerage firms, driven by greed and a willingness to gamble, jumped headfirst into the perilous business of "subprime" loans—that is, loans to borrowers with low credit scores, or exceptionally high debt-to-income ratios, or not enough money for down payments. And though in 2004 subprime borrowers were seven-and-a-half times more likely to be ninety days delinquent than "prime" borrowers were, and eight times more likely to be foreclosed, lenders gave them the money anyway. That's because the subprime loan industry was huge, having witnessed a nationwide increase, in loans outstanding, from $65 billion in 1995 to $332 billion in 2003 to $1.3 trillion in 2007.[69] Obviously, lenders embraced the industry for the money because subprime borrowers paid exceptionally high interest rates, sometimes 10 percent or more on a thirty-year mortgage when the conventional rate was 6 percent. They also paid ridiculously inflated "prepayment penalties" if they refinanced their loans and traded low mortgage payments on the front end for higher, adjustable-rate payments at the back.

But what did First Home care? If banks were willing to give the loans and buyers were willing to take them, then so be it. And besides, the Lease Purchase Investment program was wildly profitable because it allowed First Home to sell the same property twice, essentially: once to an investor, then a year later to a tenant after its value had grown. And values did grow, some 27 percent countywide in 2004, but 49.4 percent in Cape Coral and 94.9 percent in Lehigh Acres, where First Home did its business. "It's the strangest, strongest year I've ever seen," insisted Ken Wilkinson, the Lee County property appraiser, in 2005. "Five or six years ago, you couldn't give those Lehigh lots away. Now, you can't touch them for less than $30,000, and values all over the county keep going up."[70]

Yet, even casual observers could see something was amiss. In March 2004, for example, the Sunday edition of the Fort Myers *News-Press* had five different real estate sections totaling fifty-six pages but only eight pages of jobs. A Kiwanis Club in nearby Sarasota County had given up its traditional pancake breakfasts as fundraisers to instead flip houses, while developers in Miami blissfully planned one new condominium unit for every twenty-four residents.[71] "It's clearly a speculative bubble," explained Dean Baker, an economist at the Center for Economic Policy Research in Washington, D.C. "When you get home prices vastly outpacing inflation, there has to be a reversal. It's a huge distortion" and besides, "we need teachers . . . [not] more people flipping properties."[72]

One of the first big industry warnings came in August 2005, when Richard DeKaser, the chief economist of the Cleveland-based mortgage giant National City Corp., announced that out of 299 metropolitan areas that made up 80 percent of the U.S. housing market, fifty-three of them were "extremely overvalued." DeKaser described an area as overvalued if prices there were 30 percent above what he estimated they should be based on factors such as incomes, historic home prices, average interest rates, and population density. In all, eleven Florida statistical areas made the list, including Naples, which came in at number three at 62 percent above value; Port St. Lucie was number seven at 58 percent above value; Sarasota ranked twenty-six at 45 percent; Cape Coral, forty-two at 35 percent; and Deltona, forty-three at 31 percent.[73]

Then, there were the big investors, such as hedge-fund manager Michael Burry, later made famous by the Michael Lewis book and subsequent movie *The Big Short* featuring actor Christian Bale. In 2005, Burry not only predicted but actively sought to profit from a nationwide housing collapse (he guessed it would come in 2007 and made over $2.5 billion on insurance-like financial derivatives known as credit default swaps). And also in 2005, well-known Florida analyst Jack McCabe of McCabe Research & Consulting, a Deerfield Beach industry intelligence firm, began putting together what he called "property pools," small investment groups of up to twenty-five people who gathered some $10 million in capital to buy up what McCabe predicted would be thousands of distressed condos in Florida beginning in the first quarter 2006.

"When prices have been artificially increased by speculators flipping units back and forth, and always at a higher price," said McCabe, "that works fine until . . . there's nobody out there [willing] to pay a higher price than you paid." It was Florida's version of the "Greater Fool" theory, he explained, and Florida had run out of fools.[74] Although certainly not a uniform process, the slowing of home sales in the state began in the third and fourth quarters of

2005. Then, through mid-2006, sales plummeted and listings rose precipitously, but sellers held firm on the price. They were "ostrich investing," said experts, expecting high rates of return as speculators but refusing to see that huge home inventories with no buyers would someday lead to a crash. They "had this mentality of, 'I'm going to flip it and make $50,000 in six months,'" insisted West Palm Beach investment lawyer John Pankauski in 2006, and they're "just now starting to realize that not only are they not going to make that much money, they will probably lose money." But "there's a difficulty [in] admitting they made a bad choice," so right now "very few sellers are dramatically reducing their prices to unload."[75]

However, one bad quarter turned into two bad quarters; two bad quarters into a year. In Lee County, for example, year-to-year sales of existing homes in December 2006 dropped 45 percent, from 1,084 to 598, while the median price dropped 19 percent, from $322,300 to $262,500.[76] Cheerleading realtors said not to worry, that this was a "correction," a chance for the market to "rebalance itself," but by 2007 no one was selling homes, not even First Home, which had been acquired at the height of the housing bubble by New Jersey–based Hovnanian Enterprises, the sixth-largest homebuilder in the country, for $350 million in cash. "Hovnanian was a little late to the party," asserted industry expert Alex Barron of the Agency Trading Group. In August 2005, it "went from no exposure" to America's property mess to buying First Home and "having 25 percent of [its] market in Florida."[77]

It was a bad move, for not only did First Home's sales retract, massively, but existing buyers canceled their contracts in droves—a full 35 percent of them canceled in the fourth quarter of 2006. Then, in 2007, Randolph and Daphne Sewell, a working-class Miami couple who'd been hoodwinked by the First Home Lease Purchase Investment program, filed a class-action lawsuit against Hovnanian, Pat Logue, and others claiming that First Home had knowingly committed securities fraud and mortgage fraud when it breached over 600 contracts by failing to provide the Sewells and other investors with tenant-buyers and a 14 percent return. In the Sewells' case, the couple had a combined income of less than $90,000 a year, but First Home had somehow arranged mortgages for the Sewells for two houses in Lehigh Acres and one in Cape Coral for a combined price of $790,000. But, when the market collapsed and tenant-buyers failed to materialize, First Home ignored them.

"There's something desperately wrong with the whole thing," said Daphne Sewell, who admitted that without buyers, she and her husband had defaulted on two of the three loans. "I think everyone was just kind of led down the garden path."[78] The Sewells filed their suit in April 2007. By then, Hovnanian had written off more than $90 million in losses related to its Florida investments, had its debt rating lowered by Moody's Investors Service to "Ba2," which signified a "substantial credit risk," and experienced, after nine consecutive years of profits, two quarterly losses in a row. The company's First Home purchase wasn't completely to blame. After all, Hovnanian was a $4 billion company, and in March 2007 sales of existing homes had plummeted nationwide, experiencing their largest single-month drop in eighteen years. But, as Hovnanian vice president J. Larry Sorsby asserted in June, Lee County "is by far the worst housing market that we're in, and I wouldn't be surprised if it's the worst housing market in the country."[79]

The figures were alarming, even surreal. By fall 2007, First Home had a backlog of 1,787 unsold homes in Lee County worth $496 million. What's more, Hovnanian had laid off over 1,100 First Home workers and was using a fifty-person skeleton crew to liquidate its properties in bulk. Then, in August 2009, Hovnanian scuttled First Home by selling the company's remaining assets, some 1,409 lots scattered about Lehigh Acres, to a group of Czech investors for $4.2 million, or less than $3,000 apiece—roughly a tenth of what they were worth when Hovnanian purchased First Home in 2005.[80] Nevertheless, Hovnanian would weather the storm, returning to profitability in 2014, but numerous other builders, including Florida-based TOUSA of Hollywood, WCI Communities of Bonita Springs, and Levitt and Sons of Fort Lauderdale, failed.

In 1968, the corporate giant ITT had purchased Levitt & Sons to develop Palm Coast, the planned community near Jacksonville, but the once-celebrated company had passed through two different owners and several recessions before reemerging as an independent company called the Levitt Corp. in 2003. The Levitt Corp.'s homebuilding unit was a subsidiary named Levitt and Sons (written without the ampersand), which in 2005 started construction on a 1,200-home planned community in Port St. Lucie called the "Seasons at Tradition" for those age 55 and older. But, as the market turned, Levitt and Sons defaulted on more than $300 million in loans, declaring bankruptcy and leaving what the *Palm Beach Post* described as a "ghost town." "It's very scary when you feel totally abandoned," said Susan Visakowitz, a Seasons resident and retired nurse from New York who lived in a cul-de-sac full of empty homes.[81]

In all, the Seasons had just ninety homes that were finished, another sixty or so without doors or drywall, some with tiles stacked on the roof, while at most, some thirty people lived there. The reason was that a majority of those who purchased the properties were speculators; they lived elsewhere or had lost jobs and burned through their savings and their banks had foreclosed. That, perhaps, was the greatest symbol of Florida's property meltdown: the abandoned, foreclosed home. They were everywhere, as Florida trailed only California in foreclosures, with 279,325 filings in 2007 on 165,291 homes, up 200 percent from 2006, and 2 percent of all Florida properties as a whole. Essentially bank repossessions, foreclosures jumped in 2007 because property had declined in Florida in 2006. People owed too much money. They had overextended themselves by purchasing expensive homes or by speculating in homes and now owed more to lenders than their properties were worth. Their mortgages were "underwater."

Meanwhile, lenders who had given the subprime loans—worth about $1.3 trillion as of early 2007—sold them to investment banks that bundled them together with prime loans in bonds called "mortgage-backed securities," or MBSs, which the investment banks sold on Wall Street. Banks, hedge funds, wealth management funds, pension funds, mutual funds, unions, foundations, corporations, college and university endowments, local governments, state governments, and untold foreign entities all invested in MBSs, to the tune of $6.7 trillion. Thus, when the bubble burst and people began to default, investors lost big. Indeed, the world lost big—according to the International Monetary Fund, some $4.1 trillion globally in bank and other financial institution losses by April 2009.[82] By that point, the Dow Jones Industrial Average had fallen 50 percent from its 2007 high, wiping away another $11 trillion in value, while the U.S. unemployment rate jumped from an average of 5 percent in 2007 to almost 10 percent in 2009, when a then-record 20.7 million Americans collected unemployment benefits.[83]

In Florida, a state heavily dependent on real estate and construction trades, not to mention banking and tourism, things were even worse. By late 2009, fifty of the state's sixty-seven counties had unemployment rates that were higher than the national average. Overall, 11.8 percent of workers in Florida were unemployed, a staggering 1,087,000 people, which didn't include another 163,000 so-called discouraged workers who had dropped out of the labor force and no longer were looking for work, otherwise the state's real jobless rate would have been 13.4 percent, at the time the highest figure ever recorded in Florida since officials began tracking unemployment in 1970. "This [unemployment report] is bad and really it's worse than it looks because you had that drop in labor force participation," explained Scott Brown, chief economist with Raymond James Financial in St. Petersburg. "They don't call it the Great Recession for nothing."[84]

Naturally, the highest unemployment rates were in construction-dependent counties with large land giant developments, such as Flagler, where Palm Coast was located, which came in first at 16.9 percent. Hernando with Spring Hill was second at 14.9 percent; St. Lucie third at 14.2 percent; and Lee, with its Lehigh Acres and Cape Coral communities, sixth at 13.8 percent. However, Lee was a special case, where high unemployment, record home foreclosures, and thousands of bankruptcies put it first in Florida and twentieth in the United States out of more than 3,000 counties in September 2009, according to an Associated Press (AP) monthly analysis called the "Economic Stress Index." By examining data from all three categories, the AP tabulated scores of 1 to 100, with the average American county at about 10.5 and scores of 11 or more considered "stressed." Tellingly enough, Lee County's score was nearly double that at 20.9.[85]

The numbers were bad throughout Lee County, where in 2008, for example, the Cape Coral-Fort Myers metropolitan statistical area had the highest foreclosure rate in the nation, with an astonishing 41,040 properties foreclosed. In March 2009, some 55,022 Lee residents received food stamps, up 157 percent in two years, while the number of homeless families grew 115 percent in Lee between 2008 and 2009.[86] "Housing Woes Exasperate the Despair of Divorce," moaned a 2009 headline in the *News-Press*, which with a grim determination chronicled the region's collapse. Others headline read: "Suicides Intensify in Crushing Times," "Child Abuse Up in Hard Times," and "Unfinished Homes Start to Crumble."[87]

Indeed, Lee's property meltdown, in which the median price of an existing single-family home dropped 73 percent from a high of $322,300 in December 2005 to just $87,300 in April 2009, led, directly or indirectly, to a domestic violence problem. In 2008, the county saw its highest number of domestic violence killings in over fifteen years. It also saw a 67 percent jump in bankruptcy filings; a 30 percent jump in drug overdoses; a 23 percent jump in Medicaid enrollments; a 19 percent jump in uninsured hospital patients; and a 12 percent jump in the number of schoolchildren receiving free or reduced lunch. In all, more than 50,000 children, some 62 percent of all Lee public school students in 2008 and 2009, received free or reduced lunch, leading to dire warnings by the *News-Press* that without massive donations to a community-wide "Hunger Relief Fund," Lee faced a summer food shortage of what it called "natural disaster proportions."[88]

"I think people are [finally] getting it," said Al Brislain, head of the region's largest food bank, in May 2009. "The 2007 summer, I didn't get it. Housing was going bad that spring and a lot of people were getting laid off. I was new

Florida's land giant developers built an incredible backlog of cheap, affordable lots. However, during the property boom of the early 2000s, speculators drove land prices into the stratosphere. Thus, when the market collapsed, communities like Lehigh Acres were at the forefront of America's foreclosure crisis. Courtesy of Daniel Kariko.

[here] and we all thought maybe it would turn around. It didn't. Everyone gets it now."[89] In Lehigh Acres, where the value of taxable property dropped nearly $4 billion, from $7.19 billion in 2007 to $3.37 billion in 2009, and where homes routinely sold in the $45,000 range, the feeling, insisted *St. Petersburg Times* reporter Kris Hundley, was "post-apocalyptic." "Under a huge summer sky, the sprawling community appears overexposed and abandoned," she wrote. "The main drag offers an endless loop of failed businesses and vacant strip centers," while along its "cookie-cutter side streets . . . [there's an] empty lot here, [a] half-finished concrete shell of a house there, [a] rundown duplex for rent down the block. Unfinished condos have been torn down or boarded up by county crews" and "a new, eight-unit townhome project has two tenants, their homes the only ones with glass instead of wood where the windows should be."[90]

Lehigh Acres had so many abandoned houses, some with furnishings and appliances still inside, that the desolate community saw a marked increase in property crimes—51 percent more residential burglaries were reported in 2008, a startling 694, than in 2007. "Early on, I was told that subcontractors would try to get back things like sod or air conditioning units that weren't paid for," explained an official with the Lee County Department of Community Development. But "we're not seeing that so much anymore. Now we're seeing outright thievery. Someone needs a carpet, they find an abandoned house and steal the carpet. Same thing with appliances being stolen."[91] Interestingly enough, one of the biggest targets was the outside portion of people's air-conditioning units called the condenser, which thieves would rip apart for scrap metal or pick up intact and haul away in trucks.

This was no small job, like moving a refrigerator or a desk, but since condensers were worth $1,500 to $3,000 installed, criminals targeted them. "You wouldn't believe how bad [it] is," explained a Lehigh Acres landscaper named Chuck Barlow, who estimated that by 2010 nearly a third of homes in the community had had their air-condition units either pinched or destroyed. So Barlow had an idea. In 2008, he began building steel cages and other anti-theft devices to attach to condensers, pool pumps, electric meters, and other outside equipment and selling them through a business called AC Stolen Inc., with the website acstolen.com. "I use 10-to-12 gauge steel [in my cages] and bolt 'em down," he told the *Lehigh Acres Citizen*. "Go ahead and try to bend that steal. You can't [even] cut it with . . . bolt cutters." And, though Barlow was new to the business, abandoned and foreclosed home burglaries were so common in Lehigh Acres, he told the *Citizen*, that he was planning to build an entire line of *inside* security devices for refrigerators, water heaters, washers and dryers, and stoves.[92]

Unfortunately, entrepreneurs like Barlow could do little to stop Lehigh Acres' biggest criminal enterprise: marijuana grow houses, which appeared in the community in unusually high numbers even before the market had collapsed. In fact, grow houses had become common in Florida by the early 1990s, as aggressive drug interdiction efforts, including the increased use of helicopters to detect outside grow operations, such as those in Golden Gate, made it safer for criminals to grow marijuana indoors. But rather than keep plants in their own homes, they often purchased properties in sprawling planned communities like Lehigh Acres, where houses were cheap and nondescript and where the comings and goings of a neighbor, two lots down on a half-built street in a maze of streets, went unnoticed. "The house was set up just for growing marijuana," explained Lee County sheriff's lieutenant Jeff Holland in 2001, following the random discovery due to a fire of a 500-plant hydroponic grow house in Lehigh Acres, where "hired professionals perhaps" had strung up Christmas lights, installed a back deck with a jacuzzi, and outfitted the living room with a couch, a stereo, lamps, end tables, and a TV to fool anyone who happened to visit. "You could have walked around the home fifty times and . . . never known it."[93]

However, inside, in three bedrooms and a garage, workers had insulated the walls, set up electric panels for high-intensity lights and water and fertilization systems, and bypassed the home's electric meter by running a buried power line to a pole outside. For $20,000 in equipment and $100,000 for a house purchased by a suspected "straw buyer" in Miami and "rented" to immigrant Cuban growers, a criminal gang, authorities believed, had produced plants worth $500,000. And "I would bet there's a lot more of them [out there]," said Lehigh Acres assistant fire chief Henry MacArty. "They can move an operation every year, year and a half, and never be caught. They can also pay $100,000 for a house" then "walk away" because "they've made so much money they don't feel it."[94]

Of course, no one knows how many grow houses there were in Lehigh Acres, but in 2007, Lee County ranked second in the number of grow houses discovered in the state—ninety-five, with over 7,646 plants seized. And in the six-month period between October 1, 2007, and March 31, 2008, some thirty-three grow houses with over 2,000 plants were discovered just in Lehigh.[95] The numbers were disconcerting, but law enforcement officials discovered similar large operations in Palm Bay, Cape Coral, Port St. Lucie, Deltona, North Port, Port Charlotte, Spring Hill, and Golden Gate, where the connection between indoor marijuana production and the anonymousness of cookie-cutter land giant communities was clear. "Welcome to paradise—for pot growers," wrote *Florida Today*, a Brevard-based newspaper that covered Palm Bay. "Pot cultivation is booming all over Florida, but Miami-area drug criminals are [coming] to Brevard because its relatively bucolic atmosphere is perfect for setting up indoor marijuana farms. . . . Call it horticultural outsourcing sponsored by gangs from South Florida."[96]

Yet, though grow houses were a real problem in Florida, with dangerous fires and gang-related home invasions and killings even in Lehigh Acres, in 2008 and 2009 the state's biggest problem, bar none, was its crippling rate of foreclosures. Already gallingly high at nearly 280,000 in 2007, foreclosure filings jumped to 385,309 in 2008, meaning 4.5 percent of all Florida housing units, or roughly one in twenty-two, were in foreclosure.[97] The process took one to two years to settle, sometimes more, and required a hearing before a judge before a home could be repossessed. That meant hopelessly clogged courts, as well as a slow-moving tsunami of deeply discounted, bank-owned properties for sale, known as "real estate owned" properties, or REOs, which glutted the market and kept already-low prices depressed. In Lee County, for example, where in November 2008, more than 60 percent of homes being sold were REOs and where prices had dropped some 42 percent in a month, there were 4,000 houses for sale for under $100,000.

"It's like Black Friday in the real estate market [here] every day," said Fort Myers realtor Denny Grimes, but "with one exception. We're not going to run out of Game Boys."[98] In Cape Coral, there were so many foreclosed properties, an estimated 11,000, that in 2008 a local real estate agent named Marc Joseph bought an old bus, which he painted bright green with the words "Foreclosure Tours R Us" on the sides and began giving tours. "I'll tell you what you can get for $100,000 today," intoned Joseph. "You can get almost a brand new home, 4 bedrooms, 2 bathrooms, and about 1,900 square feet of living space that [in 2005 went for] $280,000. . . . And houses that were $600,000 to $800,000? Those houses are down to $200,000 or $225,000. And those are Gulf-access homes. . . . [They go] straight out to the Gulf." To prove his point, in 2009 Joseph added a pontoon boat to his fleet, which took northerners and Canadians and even some Europeans on a canal-front tour of homes. He was providing a service, he said, because foreclosed homes distorted the market.[99]

Sold "as-is" and often at a loss by banks in so-called short sales, these houses kept prices low, which ate up equity, limited the ability of even solvent owners to move because their homes were "underwater," and killed the construction industry, which was a leading employer in Lee that pulled some 9,747 single-family home permits in 2005, but just 331 in 2009.[100] "I would say there's about three years of inventory to work through" in Cape Coral, explained Russell Weyer, an associate with the Naples firm Fishkind and Associates, so "if you're a builder, you're going to wait this one out. . . . [These houses] won't get absorbed overnight."[101] Although banks often sold properties in bulk, to investors, the pace of REO home sales depended, for the most part, on the courts, for the longer judges took to hear cases, the longer rock-bottom home prices would last, at least in Lee, which had a backlog in 2008 of over 24,000 filings.

Therefore, in December 2008, Lee County Clerk of Court Charlie Green spearheaded an effort by two circuit court judges and four retired judges to implement what critics and supporters alike called the "rocket docket," a system of lightning-fast foreclosure hearings in which cases were heard in about thirty seconds. "I went to the judges and said 'We're going to have one hell of a problem if we don't get moving,'" explained Green, because, with such a huge backlog of cases, "you have all this indecision. You have all this uncertainty [and] nothing good comes from uncertainty."[102] With martial-like precision, judges would ask owners if (a) they were living in the home and (b) if they were current on their mortgage, then they issued judgment after judgment, each rapid-fire, boom-boom-boom with a signature, telling the owners to leave. If they lived in the home, they had sixty days to vacate; if not, forty-five.

On rare occasions owners contested the rulings—on Thursday, 4 December 2008, for example, the court heard 841 cases and signed foreclosure orders on 825—but judges rarely, if ever, required mediation between banks and owners to modify loan terms to actually keep people in their homes. That wasn't the goal, for at least one Lee judge was so hell-bent on clearing cases that he exempted banks from submitting legally required verifications confirming that the loan documents he had been given were valid.[103] As a result, Lee's rocket docket kept on rocketing on, making national news and drawing lawsuits and forming a contradictory backdrop to a $50 billion to $100 billion loan-modification plan for homeowners facing foreclosure then being proposed as part of a larger stimulus package by President Barack Obama, who visited Fort Myers for a town hall meeting in February 2009.

Inaugurated that January, Obama had inherited a subprime mortgage crisis that had grown into a global recession, and he visited Fort Myers as part of broader public relations effort to sell his plan. The trip was a short one; Obama spoke and took questions for maybe an hour, delighting locals when he hugged a sixty-one-year-old Lehigh Acres woman named Henrietta Hughes who not only lost her home but spent the last six months living with her adult son in her car. And, though the president didn't tour Lehigh Acres, the press did, most notably George Packer of the *New Yorker*, who, in an article titled "The Ponzi State," described Lehigh Acres as one of many inland subdivisions in Florida where trash filled the side streets and it was "possible to see" America's "slums of the future." Packer then spoke to Doug Bennett, a bureau chief at the *St. Petersburg Times*, who said glumly: "[They're] stucco ghettos. I live in one, so I can say that. Too many houses, not enough water, the economy's terrible, no tourists. This is the capital of the low-wage jobs, and when things go bad people have no safety net. . . . This is the epicenter of everything that's bad in America."[104]

# Epilogue

## *The Last of the Best of Florida*

> We're looking at this in the context of the land and not imposing a predetermined plan on the land. It's the concept of the old English village. You design a village street plan that creates a neighborhood with natural interaction.
>
> —BILL HAMMON, "Green Team" Director, Babcock Ranch

Babcock Ranch, Florida. At 91,362 acres, this enormous spread of prairie, swamp, and wet-pine wilderness was for over ninety years a continually functioning business, founded in 1914 by millionaire lumber magnate and Pittsburgh mayor E. V. Babcock, and passed to his son Fred, whose Babcock Florida Company handled its affairs. It spanned two counties: southeastern Charlotte County, where the bulk of its acreage was, and northwestern Lee County, just north of Caloosahatchee River. Twenty-four miles to Fort Myers and twenty-five miles to Punta Gorda, the ranch was extremely remote but, at the height of its operations, employed perhaps a hundred people. It had fifteen full-time cowboys, for example, who oversaw thousands of range cows, a buffalo herd, and some quarter horses, and who, as late the 1980s, maintained armed patrols to ward of cattle rustlers. It had forestry experts and tree cutters who loaded logs of termite-resistant pitch pine onto railcars on the ranch's own rail spur for export to South Africa, where the hardy wood worked especially well in diamond mines.[1]

The ranch had alligator skinners and ostrich herders and vegetable farmers and limestone miners and even children, whose parents lived and worked at the ranch but who sent them to school in Punta Gorda. The problem was that the mile-long road to the bus stop, on State Road 31, was lined on either side with retaining ponds filled with alligators, so in a pinch, one enterprising father taught his elementary-school-aged son to drive, and each morning the boy loaded younger kids into the car, started it up, and ferried them past the alligators to the fence line, where they parked and met the bus. Such was life at Babcock Ranch, where owner Fred Babcock—whose millionaire father sent him to South Dakota when he was twelve to learn the cattle business, then to sawmills in Pennsylvania to learn the lumber business before sending him to Dartmouth—was, by all accounts, a careful steward of the land.[2]

He improved the ranch's pastures by planting new and hardier grasses; experimented with crops, which he rotated; cut and replanted its timber several times; and imported Brahma, Hereford, Angus, and Charlaise bulls to replace its range cattle. Babcock also sold an adjoining, 19,000-acre tract for a wildlife management district, which still exists today, and left virtually untouched a 10,000-acre swamp at the heart of his property. However, Babcock died in 1997 at age 83, leaving four daughters and some thirty-five other heirs who wanted to develop the land. They tried, at first, to dedicate and/or sell half of their holdings to the state's Florida Forever land acquisition program and various other entities while retaining ownership to a smaller 28,000-acre ranch. In exchange, the Babcocks would receive the right to develop an additional 20,000-acre parcel on the border of Charlotte and Lee counties into a planned community the size of Lehigh Acres.[3]

Conservationists were thrilled but also appalled because the Babcock bargain was a Faustian one: priceless wetlands—which, when added to conservation areas to the east and west of the ranch, would form an unbroken, environmentally protected corridor from Lake Okeechobee to Charlotte Harbor—versus 20,000 acres of new sprawl. Instead of a housing density of one unit per ten acres, which is what Babcock Ranch was zoned for, the new Babcock community would have one unit per acre, better than Port Charlotte, to be sure, but still a dense development of apartments, strip malls, office parks, shopping centers, and homes.[4] Therefore, in 2003, by a vote of three to two, the Charlotte County Board of Commissioners demurred, citing the sheer irresponsibleness of building a town in such a remote location. It named utilities, schools, school transportation, and traffic as just a few of the problems and worried what would happen if suddenly "50,000 people . . . [fell] from the sky."[5]

And besides, the county board didn't have a land acquisition fund like the state did, and its focus was on pushing growth to northern and western parts of the county, to places like Port Charlotte, where the moribund General Development Corporation (GDC) had left the 1950s-era community with thousands of empty lots. But, in 2004, representatives from some forty national nonprofit groups—including the Conservancy, the Audubon Society, and the Trust for Public Lands, as well as state, regional, and federal officials, and commissioners from Charlotte and Lee Counties—met for a "creative thinking" session at Babcock Ranch. There was interest in and a desire to buy the property outright, all 91,362 acres of it, for as the president of the Florida

Wildlife Federation put it: "It's almost a no-brainer. This is the last of the best of Florida."[6] The obstacle, unfortunately, was the price: between $200 million and $1 billion, too rich for any one entity at the meeting, but doable, perhaps, for the group.

Therefore, the nonprofits in attendance quickly formed the Babcock Preservation Partnership to kick off a fundraising campaign while commissioners in Lee County pledged $40 million to offset the purchase price.[7] In neighboring Charlotte, though, where voluntary tax increases were almost always a no-go, officials didn't have a workable plan. They wanted to pass a 0.05 mill property tax amounting to five cents on every $1,000 in property value, a move that would raise at least $50 million over ten years. But for that they would need a yes vote from the public, and yet they hadn't told the public why Babcock was so important. Most residents had never heard of "green infrastructure" or wetlands restoration programs, but they hated taxes so much they wouldn't have agreed to the increase anyway. "The problem is, Six-Pack Charlie doesn't give a rip about the transference of carbon," explained one observer. "All he wants to know is] what he's going to get out of it."[8]

Soon, though, the issue was moot. In August 2004, Charlotte bore the brunt of Hurricane Charley, a Category 4 storm that leveled parts of both Punta Gorda and Port Charlotte and that caused some $16 billion in damage. With bigger fish to fry and hundreds of homeless residents stuck living at Punta Gorda's airport in a tent city, Charlotte was out. Then, in December, the group's main and most important member, the deep-pocketed Florida Department of Environmental Protection, which oversaw the state's Florida Forever program, offered to buy the ranch for $400 million. It was a low-ball offer, and the Babcocks had said they would only be willing to the accept the mid-price of two independent appraisals, which had come in at $450 million and $460 million, respectively. To be fair, there were tax issues and eminent domain issues that helped to torpedo the deal, but there were indications that Florida Governor Jeb Bush's pro-business administration wanted land development as part of the plan.[9]

So the Babcocks went elsewhere, to a private buyer named Syd Kitson, a former pro football player and developer of golf courses based in West Palm Beach. In July 2005, at the height of Florida's housing boom, Kitson offered the Babcocks $700 million, though under two conditions: First, Charlotte and Lee counties would allow him to build a 50,000-resident planned community on 17,880 acres at the south end of the property called "Babcock Ranch," while second, the state of Florida and Lee county would pay him $351.5 million for the rest, in all some 73,471 acres. Kitson pitched it as a win-win. So did

Charlotte's commissioners, who had come to the conclusion that development there was inevitable and thus voted four to one to change the county's comprehensive land-use plan to include Babcock Ranch. Not everyone was thrilled, especially those who lived near the ranch and staffers at the county's Community Development Department, who found no fewer than thirty-six "issues" with Kitson's application and who feared that approving the community would create additional sprawl.

"We've heard these promises before," said commissioner Adam Cummings, who voted against the project, "and look at how that turned out. [You'd think we would] learn from history."[10] Nevertheless, in July 2006, Kitson closed the deal. He planned to build his new community beginning in 2008 and shocked observers by promising a "green" city, one with clustered home development, 10,000 permanently protected acres of parks, wildlife buffers, flow-ways, and energy-neutral buildings powered by a seventy-five-megawatt generator on a ranch-based, 350-acre solar farm.[11] He pitched it as "the first solar-powered city in America" and insisted that unlike Port Charlotte, a community caught in a "straitjacket of uniformly sized parcels" with "piecemeal pockets of car-dependent homes," Babcock Ranch would have a "vernacular" focus, in which the ebb and flow of the landscape, and not the developer and his bulldozers, would determine its design.

"We're looking at this in the context of the land," said environmental sciences professor and Kitson "green team" director Bill Hammon, "and not imposing a predetermined plan on the land. It's the concept of the old English village. You design a village street plan that creates a neighborhood with natural interaction."[12] Thus, instead of thousands of lots on a gridiron lined with strip malls, Babcock Ranch would have two villages and five hamlets flowing to a city core. There'd be forty-two miles of hiking trails woven through the development and electric shuttles. Houses built according to Florida Green Building Coalition standards and designed with deep porches in various pre–World War II styles would be set close to sidewalks—to, as Kitson's company put it, "encourage conversations with passersby and allow homes to live from front to back."[13] There'd be a "Founder's Square" with a band shell, a civic center, and a lakefront green with a promenade and paddleboard rentals; a splash pad for kids; land for schools and parks; and 6 million square feet of commercial space, with a focus on walkability.

In short, Babcock Ranch was the opposite of Port Charlotte and was constructed as an ecofriendly, "New Urbanist" take on community design. In Florida, it's a type of community that's proved increasingly popular with, for example, New Urbanist developments at Celebration, Reunion, Harmony,

Avalon Park, Baldwin Park, and Independence near Orlando; Haile Plantation near Gainesville; Abacoa near Jupiter; and Seaside in the Panhandle, which in 1998 was the picture-perfect though somewhat creepy setting of the Jim Carrey film *The Truman Show*. There are even New Urbanist communities in Southwest Florida in addition to Babcock Ranch, including Ave Maria near Naples, Palmer Ranch near Sarasota, and the West Villages in North Port, while Port Charlotte, Cape Coral, and Lehigh Acres are all attempting to reenergize their communities using New Urbanist designs.

It's been slow-going. First, the mortgage crisis of 2007 and 2008 caused a recession in Florida that left an enormous backlog of homes, including tens of thousands of homes in Lee and Charlotte Counties. Even Babcock Ranch, which had the financial backing of Morgan Stanley, missed its 2008 opening date, then its 2011 opening date, and wouldn't sell homes until 2017. Then, in gridlocked communities such as Port Charlotte, people have found that assembling large parcels from many small parcels zoned single-family residential has been almost impossible to do. Charlotte County tried it in the early 2000s by eminent-domaining some seventy-five homes, sixteen businesses, three duplexes, three churches, and a whopping 3,000 empty lots in a barren section of northern Port Charlotte in Murdock, where the county hoped to build a mixed-use town center with restaurants, shops, a walking district, government and community buildings, and condominiums.

The plan was to offer owners full market value for their properties, about $2,000 to $3,000 a lot, though some, duped by GDC and living as far away as Singapore, had paid many times that. "[You'd] be surprised at some of the prices some of these people paid for these parcels," said J. Paul Payette, manager of the county's real property services, sometimes "$18,000 to $20,000 for [lots] . . . assessed at $2,000 or $3,000. So guess what happens when we offer them that?"[14] The answer: they sued. For one, they wanted more money, but additionally the county hadn't condemned the property for "public use," which is what eminent domain is really for, at least in theory—the taking of private farmland for a reservoir, for example, or houses for an electric plant—but instead for the "public good." And no doubt, fixing Port Charlotte's prematurely platted grid system was a public good. The community needed a downtown, desperately. But the county planned to seize the property through eminent domain then flip it to a private developer for private use.

Thankfully, the state would kill the practice in 2006, but not before Charlotte officials acquired a 1,132-acre parcel at a cost of $93 million. Faced with interest payments of $14,300 a day, they needed to sell the property quickly, but when the recession hit, several interested developers, including Syd

Kitson, backed out.[15] Thus, in a boondoggle for the ages, what began as a wildly expensive eminent domain project sat for over a decade then turned into a planned development of 2,000 single-family homes, plus a water park, and "a prime music and entertainment destination" called Arredondo Pointe, which, though set to feature an amphitheater and several large hotels, will not be the New Urbanist downtown they had hoped for.

Instead, it's just sprawl, more sprawl in a sea of sprawl that's the 1950s-era planned community of Port Charlotte. Today, Florida has many Port Charlottes. They are the platted remnants of an earlier age in which lot sales trumped home sales, intelligent planning was an afterthought, and salesmen, hucksters, con men, and corporations subdivided then sold the state to a generation of retirees. And now, with COVID-19 pushing more and more people to the suburbs, to cheap homes on cheap lots in similar communities throughout the state, houses on those rows and rows of homesites, those barren lots on "grassphalt" roads that once led nowhere, are finally being built.

# *Acknowledgments*

A great number of people were instrumental in helping me conceive and/or research this book, which grew out of my own lifelong interest in the history and culture of Punta Gorda, Florida, and its environs. I was raised there, in a 100-year-old Victorian home on a brick street lined with palm trees leading to a park on Charlotte Harbor. It wasn't St. Augustine by any means, but Punta Gorda was old for Florida and contrasted sharply with the communities around it, in particular its sister-community Port Charlotte. For whereas Punta Gorda was a product of the late nineteenth century and once was a violent frontier town full of saloons and brothels, Port Charlotte was an assembly-line product of the twentieth century, a planned community built for retirees. Their origins were entirely different. So, that's my first acknowledgment. I owe the idea for this book to the twin yet very distinct communities of Punta Gorda and Port Charlotte. They were, to me, the old Florida and the new, and were an interesting place to grow up.

Then, there are the people who have helped me: friends, family members, acquaintances, writers, editors, academics, land developers, land planners, and archivists, among others, whose input, offers of help, and/or permissions, both big and small, were instrumental to this book. In no particular order, they are: Janel Trull of the Cape Coral Historical Museum; Jennifer M. Zoebelein and Crystal Diff of the Charlotte County History Services; Heather Otis of the Marco Island Historical Society; the staff at the State Archives of Florida; members of the DFW Writers Workshop; and Joel Desguin, Vic Desguin, Binky Waldrop, Rex Goulding, Charlie Telfair, Al DuBoice, Wendy Weaver Wilhelm, Josh Christensen, Dennis Westby, Joanna Viarengo Weaver, Robert McQueen, Paula McQueen, Tom Frame, Lynn Harrell, Debbie Aughey Barrera, Ted Aughey, Bill Marcum, Craig Pittman, Gary Mormino, Canter Brown Jr., Guy Batsel, Rod Linder, Darin Hinshaw, Wally Hinshaw, Wesley Wankelman, Warren Wankelman, Paul DeGaeta, Trevor Bicking, Jody Bicking, Mike Polk, Paul Polk, Travis Puterbaugh, Jeff Glas, Derek Zimmer, Greg Cantrell, Todd Moye, David Farber, Mary Anderson, Libby Bessire, David Bessire, Kevin Dixon, Sarah Dixon, Alton Peacock, Matt Boda, Estus Whitfield, Bill Stroud, Bill Spikowski, Wayne Daltry, Robert Kerstein, Amy Dare-Reeves, Erin Greb, Max Forgey, Jim Beever, Jeff Hartdorn, Steve Fasnacht, Daniel Kariko, Kevin Borland, Rob Walsh, Frank Mackle III, Dylan White, Debbie Gershenowitz, Kurt Piehler, Susan Contente, Nathaniel P. Reed, Joseph Prestia, Sue Coppernoll, Lisa Ramirez, my parents Nancy and Gerby Vuic, my sister Kelly Vuic Oliva, Mila Oliva, Angel Oliva III, Angel Oliva IV, and finally, my talented, beautiful, and supremely patient wife Kara Dixon Vuic and our delightful children Asher and Imogene. Thank you.

# Notes

## Introduction

1. Elizabeth Whitney, "Soggy Land Slips through Loopholes," *St. Petersburg Times,* 22 March 1970; Elizabeth Whitney, "Hurry, You Can't Lose (or Can You?)," *St. Petersburg Times,* 3 May 1970.

2. Elizabeth Whitney, "Swamp Sales to Out-of-Towners Was Story of a Lifetime," *St. Petersburg Times,* 18 February 1990.

3. "Levittown: The Prototypical American Suburb," *The Guardian,* 28 April 2015.

4. John Rothchild, *Up for Grabs: A Trip through Time and Space in the Sunshine State* (New York: Viking, 1985), 91.

5. "Lure of Florida Real Estate Sweeps over Entire Country," *Asbury Park Press* (NJ), 5 July 1959.

6. Seth Lubove, "Developers in Decline," *Miami Herald,* 28 May 1984.

7. Elizabeth Whitney, "Big Names Help Push Land Sales," *St. Petersburg Times,* 5 May 1970.

8. Jacquin Sanders, "Man Is Turning Manatee Counting into a Challenge," *St. Petersburg Times,* 14 February 1985.

9. Hubert B. Stroud and William M. Spikowski, "Planning in the Wake of Florida Land Scams," *Journal of Planning Education and Research* 19 (1999): 34.

10. Roger Williams, "The Politics of Planning," *Punta Gorda-Port Charlotte Florida Weekly,* 16 November 2017.

11. Robert D. Shaw Jr., "State Dries Out Swamp Peddlers," *St. Petersburg Times,* 21 October 1973.

## Chapter One

1. Frederic Remington, "Cracker Cowboys of Florida," *Harper's* (August 1895): 339–45.

2. Gene M. Burnett, *Florida's Past: People and Events That Shaped the State,* vol. 3 (Sarasota: Pineapple Press, 1991), 103.

3. Brady Vogt, "The Kings of the Cracker Cowboys," *Heartland* (June 2015): 35.

4. Clifton Johnson, *Highways and Byways of Florida* (New York: Macmillan, 1918), 213.

5. Lindsey Williams and U. S. Cleveland, *Our Fascinating Past: Charlotte Harbor: The Early Years* (Punta Gorda: Charlotte Harbor Area Historical Society, 1993), 352–53.

6. "Looking for Land?" *Desoto County News* (FL), 14 January 1915.

7. Williams and Cleveland, *Our Fascinating Past,* 360.

8. "Florida Coast Land Deal On," *Winnipeg Free Press,* 5 June 1954; Williams and Cleveland, *Our Fascinating Past,* 362.

9. "Clewiston Is Going Forward," *Chicago Daily Tribune,* 10 February 1926.

10. Frederick Lewis Allen, *Only Yesterday: An Informal History of the 1920s* (1931, repr., New York: Harper Perennial, 2010), 241.

11. Paul S. George, "Brokers, Binders, and Builders: Greater Miami's Boom of the Mid-1920s," *Florida Historical Quarterly* 65, no. 1 (July 1986): 35, 41–42; Tina Bucuvalas, Peggy A. Bulger, and Stetson Kennedy, *South Florida Folklife* (Jackson: University Press of Mississippi, 1994), 205.

12. Morton C. Paulson, *The Great Land Hustle* (Chicago: Henry Regnery, 1972), 46.

13. James Thorner, "Pinellas Failed to Learn Real Estate Lessons of 1920s," *St. Petersburg Times*, 5 July 2008.

14. Jenny Staletovich, "Ninety Years Later, the Great Miami Hurricane a Scary, 'What If?'" *Miami Herald*, 28 May 2016.

15. Ben Montgomery, "Changing Winds Toppled Castles in the Air," *St. Petersburg Times*, 26 January 2007.

16. Mary Evertz et al., "The Day a Nation Crashed," *St. Petersburg Times*, 29 October 1989.

17. Anthony Atwood, "A State of War: Florida from 1939 to 1945" (Ph.D. dissertation, Florida International University, 2012), 24.

18. "Miami Beach Wartime Training Center," *Miami Herald: Flashback Miami*, accessed 27 July 2020, http://flashbackmiami.com/2015/02/25/miami-beach-wartime-training-center/.

19. "Miami Spectacle," *Life*, 21 February 1944, 40–41.

20. "Miami Worries about Another Boom," *Life*, 12 February 1945, 65.

21. "Sergeant Rates Miami No. 1 Town," *Miami Herald*, 7 June 1945.

22. "Homecoming for Veterans," *Tuscaloosa News*, 27 February 1945; Jon L. Wilson, "Shaping the Dream: A Survey of Post–World War II St. Petersburg, 1946–1963" (master's thesis, University of South Florida, 2009), 4, 58.

23. "Hinterland Boom Grows in Florida," *New York Times*, 26 August 1956.

24. "Big Everglades Land Deal Made," *News-Tribune* (FL), 5 October 1953.

25. Vince Conboy, *Exposé: Florida's Billion Dollar Land Fraud* (Naples, FL: self-pub., 1972), 58.

26. Louis Brancaccio, "Citizens Decrying PC Road Situation," *Daily Herald-News* (FL), 11 July 1973.

27. David E. Dodrill, *Selling the Dream: The Gulf American Corporation and the Building of Cape Coral, Florida* (Tuscaloosa: University of Alabama Press, 1993), 42.

28. Dodrill, *Selling the Dream*, 25.

29. "Biggest Land Deal Recorded," *Punta Gorda Herald*, 3 November 1955.

30. "Big Sellers of Sunshine," *Life*, 9 February 1959, 77.

31. "Mothers Showing Enthusiasm in 'Better Babies' Contest," *Atlanta Constitution*, 29 October 1915; "Charlie Cox Wins Mackle Tourney," *Atlanta Constitution*, 8 August 1915.

32. Tom McEwen, "Frank Mackle, Jr." *Tampa Tribune*, 13 July 1992.

33. McEwen, "Frank Mackle, Jr."

34. Paul Lomartire, "Port St. Lucie—How the City Began," *Palm Beach Post*, 28 July 2007.

35. Frank Mackle III, "Post-War Home Construction," The Mackle Company, accessed 10 February 2017, http://www.themacklecompany.com/femjrstorypublic/09-keybiscayne.htm.

36. Joan Gill Blank, *Key Biscayne: A History of Miami's Tropical Island and the Cape Florida Lighthouse* (Sarasota, FL: Pineapple Press, 1996), 162.

37. Bea Moss, "Key Biscayne: The Natives Return," *Miami Herald*, 14 July 1985.

38. "City Group Moves to Aid Growth Here," *Punta Gorda Herald*, 30 September 1954; "Maule Group Get Welcome at Barbecue," *Punta Gorda Herald*, 14 October 1954.

39. Frank Mackle III, interview with author, 19 January 2017.

40. McEwen, "Frank Mackle, Jr." See also Gary Mormino, *Land of Sunshine, State of Dreams: A Social History of Modern Florida* (Gainesville: University of Florida Press, 2005), 54–55.

41. "Mackle Opens Ad Campaign," *Punta Gorda Herald*, 3 October 1957.

42. "Port Charlotte: The Future Belongs to Those Who Plan Ahead," *Life*, 29 September 1958, 76–77.

43. Greg Martin, "$10 Down, $10 a Month: The Sad, Scandalous Story of How General Development Corp. Sold Southwest Florida," *Sarasota Magazine*, Summer 1990, 34; Frank Mackle III, "A New Business," The Mackle Company, accessed 10 February 2017, http://www.themacklecompany.com/femjrstorypublic/10-generaldevelopment1954-1957.htm.

44. Stan Johnson, "Development Corp. Reports Increase of 213%," *Miami Herald*, 13 March 1959.

45. Mackle, "A New Business."

46. "The Brothers Mackle and General Development," *Florida Trend* (December 1958): 27.

47. Robert "Bucky" McQueen, interview with author, 7 January 2017.

48. "General Development," *New York Times*, 10 March 1961.

49. "Put Mackle Home in Chicago Store," *News Tribune* (FL), 17 January 1960.

50. Liliane Parbot-Johnson, "One of First Port Charlotte Residents Dies," *Sun Herald* (FL), 20 July 2000.

51. Jody Bicking, interview with author, 18 November 2016.

52. GDC's Port Charlotte insert appeared in both the *New York Times* and the *Chicago Tribune* on 13 March 1960.

53. "The Brothers Mackle and General Development," 27. See also Martin Abramson, "A New Life—at a Leisurely Pace," *Today's Health* 36 (December 1958): 23.

54. "See How Folks Live at Port Charlotte," *Chicago Tribune*, 13 March 1960, advertising insert, 9.

55. "Big Sellers of Sunshine," 80.

56. "No Street Lights," letter to the editor, *News-Press* (FL), 6 July 1962.

57. Robert Caro, "Some Retiree Cities Work Well," *St. Petersburg Times*, 14 August 1963.

58. Ben Funk, "Mackle Boys Build Homes for Retirees," *Muncie Star*, 25 May 1958.

59. Robert W. Delaney, "Retirement Villages Making Florida Haven for Old Folks," *Miami News*, 20 April 1958.

60. "Pensions in 1961 Total $21 Million," *Mansfield News-Journal*, 21 March 1962; R. O. Beckman, "Florida Leads in Popularity with Oldsters," *Fort Lauderdale News*, 7 July 1958; "50,000 Retired Military Personnel in Florida," *St. Louis Post-Dispatch*, 3 June 1965.

61. Burt Garnett, "Florida Census in 1970 May Hit 1½ Million," *St. Petersburg Times*, 1 October 1965; Abramson, "A New Life—at a Leisurely Pace," 23.

62. Lizabeth Cohen, *A Consumers' Republic: The Politics of Mass Consumption in Postwar America* (New York: Vintage Books, 2004), 195.

63. Cohen, *A Consumers' Republic*, 203–4; Herbert J. Gans, *The Levittowners: Ways of Life and Politics in a New Suburban Community* (New York: Vintage Books, 1967), 34. The $5,500 a year figure refers to salaries in Levittown, New Jersey in 1958.

64. "Political Sinecures," letter to the editor, *News-Press* (FL), 25 April 1962.

65. "Big Sellers of Sunshine," 81.

66. Rodney Crouther, "Port Charlotte's Sales Force Reunites," *Sarasota Herald-Tribune*, 30 May 1997.

67. Jack V. Fox, "Retired Man Finds Place in Florida," *Cumberland Daily Times* (MD), 27 August 1961.

68. Barbara Huff, "Cultural Center of Charlotte County: The Place That Friendship Built," *Charlotte Sun* (FL), 17 March 2013.

69. Anthony Wolff, *Unreal Estate: The Lowdown on Land Hustling* (New York: Sierra Club, 1973), 187–88.

70. Leslie Allan, Beryl Kuder, and Sarah L. Oakes, *Subdivisions in Florida's Wetlands*, vol. 2, *Promised Lands* (New York: INFORM, 1977), 99.

71. "North Port," *Tampa Tribune*, 3 May 1981.

72. Ben Funk, "A Real Pro Puts Mackle on the Map," *Miami Herald*, 15 February 1959; Lindsey Williams, "Toledo Blade Road Name Curious to Winter Visitors," *Charlotte Sun* (FL), 10 March 2002.

73. John Baker, "Smuggle Suspect Identified," *Daily-Herald News* (FL), 12 January 1974; Paul Marose, "Police Find Truck Suspected for Pot," *Daily-Herald News* (FL), 26 October 1977.

74. Gary Roberts, "Charlotte County Gives Failing Grade to Some—Local Roads," *Charlotte Sun* (FL), 21 August 2015.

75. Paulson, *The Great Land Hustle*, 66.

76. Wolff, *Unreal Estate*, 248.

77. Haines Colbert, "Construction Follows Wild Selling Spree," *Miami Herald*, 20 April 1964.

78. Sam Frear, "Subdividing the Wide Open Spaces," *Eugene Register-Guard*, 12 September 1965.

79. "Phony Salesmen Claimed Making Millions in Giant Land Scandal," *Lubbock Avalanche-Journal*, 26 October 1962.

80. "Cowles Buying Out Mackle Brothers," *News Tribune* (FL), 8 February 1962; Pete Weitzel, "Mackles Merge with CKP Firm," *Miami Herald*, 18 December 1962.

## Chapter Two

1. "Nation Joins in Epitaph for Connie Mack," *Santa Cruz Sentinel* (CA), 9 February 1956.

2. Eileen Bernard, *Lies That Came True: Tall Tales and Hard Sales and the Making of Cape Coral, Fla.* (Fort Myers: Press Printing, 1991), 60.

3. David E. Dodrill, *Selling the Dream: The Gulf American Corporation and the Building of Cape Coral, Florida* (Tuscaloosa: University of Alabama Press, 1993), 17. See also Bernice Freiberg, interview by David E. Dodrill, Lee County Oral History Collection, University of Florida Digital Collections, 31 January 1988, 11.

4. Ed McMahon and David C. Fisher, *When Television Was Young* (Nashville: Thomas Nelson, 2007), 277.

5. "Just One Minute, Young Lady!" *Life*, 6 October 1952, 84.

6. Dodrill, *Selling the Dream*, 24.

7. Freiberg, interview by Dodrill, 11.

8. Dodrill, *Selling the Dream*, 25.

9. "Another Heights," *News-Press* (FL), 1 September 1955.

10. "Financial Newspaper Takes a Close Look at Real Estate Promotions in Florida," *Tampa Tribune*, 27 July 1958; "Lots under Water in Harbour Heights, Widow Tells Court," *News-Press* (FL), 11 August 1962; "2,257-Acre Development Progresses," *St. Petersburg Times*, 2 October 1956.

11. Kenneth Schwartz, interview by David E. Dodrill, Lee County Oral History Collection, University of Florida Digital Collections, 16 November 1987, 21; "Filed Suit Claims Land Underwater," *St. Petersburg Times*, 12 August 1962.

12. Schwartz, interview by Dodrill.

13. Robert Caro, "Land-by-Mail Is Harbour Heights Heartbreak," *St. Petersburg Times*, 13 August 1963; "Mendelsohn Pleads Guilty to Swindle," *Cumberland News* (MD), 5 May 1970.

14. "Thomas to Sign Bond," *News-Press* (FL), 15 October 1931.

15. "Strict Enforcement in Lee," *News-Press* (FL), 17 October 1931.

16. Craig Pittman, *Oh, Florida!: How America's Weirdest State Influences the Rest of the Country* (New York: St. Martin's, 2016), 41.

17. Jim Norman, "Bay Waterfront: A Woeful Waste," *Pensacola News-Journal*, 27 October 1974.

18. Leslie Allan, Beryl Kuder, and Sarah L. Oakes, *Subdivisions in Florida's Wetlands*, vol. 2, *Promised Lands* (New York: INFORM, 1977), 142.

19. Bernard, *Lies That Came True*, 47.

20. "Florida Wrestles Land Boom" *Chicago Tribune*, 1 May 1966.

21. George Eberl, "Florida: The Sales Lures," *Stars and Stripes*, 12 November 1968.

22. Ted Shoemaker, "Florida Real Estate Booms in Europe," *Florida Trend*, August 1961, 21. As to bird-dogging, see George Eberl, "Land Sales to Military in Europe $30 Million Annually," *Stars and Stripes*, 12 December 1972.

23. Tom Hamrick, "Two Hours from Harriet's Knee," *Pacific Stars and Stripes*, 1 October 1972; Anthony Wolff, *Unreal Estate: The Lowdown on Land Hustling* (New York: Sierra Club, 1973), 14–15.

24. Joe Raso, interview by David E. Dodrill, Lee County Oral History Collection, University of Florida Digital Collections, 7 May 1988, 6.

25. "Europeans See Florida as Ideal Place to Invest," *Orlando Sentinel,* 14 March 1965; Roy Bertke, "Foreigners Eye Land in Florida," *Tampa Times-Tribune,* 30 September 1973.

26. Bernard, *Lies That Came True,* 98.

27. "Subdivision Promotion," *News-Press* (FL), 5 March 1955.

28. Kris Hundley, "Lehigh Acres: Florida's Lesson in Unregulated Growth," *Tampa Bay Times,* 8 August 2009.

29. Paul Reyes, *Exiles in Eden: Life among the Ruins of Florida's Great Recession* (New York: Henry Holt, 2010), 210.

30. Bernard, *Lies That Came True,* 114, 116.

31. "Go A.W.U.T.F.," *Daily News* (VA), 30 June 1968.

32. Talis Bergmanis, "'Free Trip' to Florida Has Joker," *Democrat and Chronical* (NY), 19 January 1968.

33. "State Vacation Gimmick Flayed by Land Board," *News-Press* (FL), 11 July 1968.

34. "Florida Hits 'Free Trip' Land Deals," *Chicago Tribune,* 14 September 1969.

35. Elizabeth Whitney, "Lovely Homesites," *Minneapolis Star,* 1 February 1972.

36. Elizabeth Whitney, "Top Secret: The Script That Isn't," *St. Petersburg Times,* 4 January 1971.

37. Bernard, *Lies That Came True,* 63.

38. Morton Paulson, "Big Sell Follows Free Dinner," *Daytona Beach News-Journal,* 25 April 1965.

39. Elizabeth Whitney, "Wine 'em, Dine 'em, Then Sock It to 'em," *St. Petersburg Times,* 5 January 1971.

40. Richard E Rustin, "Land Sales Toss Parties to Push Sales of Homes and Plots," *Wall Street Journal,* 15 March 1966.

41. Dodrill, *Selling the Dream,* 74.

42. Fred Farris, "Harbour Heights Firms Goes under Bankruptcy Act," *News-Press* (FL), 2 October 1962.

43. Whitney, "Wine 'em, Dine 'em, Then Sock It to 'em."

44. Joe and Agnes Miller, interview by David E. Dodrill, Lee County Oral History Collection, University of Florida Digital Collections, 27 January 1988, 11.

45. Bernard, *Lies That Came True,* 103.

46. Lynn Harrell, email correspondence with author, 11 May 2017.

47. Bernard, *Lies That Came True,* 103.

48. Bernard, 93–94.

49. Jack Nease, "Developer Appears to Whitewash Filing," *Sun-Sentinel* (FL), 22 March 1990.

50. Leonard Rich, interview by David E. Dodrill, Lee County Oral History Collection, University of Florida Digital Collections, 27 January 1988, 38.

51. Jody Bicking, interview with author, 18 November 2016.

52. United State Census Bureau, *1970 Census of Population. Volume 1. Characteristics of the Population, Part 11: Florida* (Washington, DC: U.S. Government Printing Office, 1973), 100, 116, 140.

53. "Meet a Real Estate Tycoon," *News-Press* (FL), 18 April 1965; "58—Lots for Sale," *News-Press* (FL), 22 February 1965 and 24 May 1965.

54. Rick Berry, "The Shifting Sands of Florida Buys," *Tampa Tribune-Times*, 3 May 1981.

55. Berry.

56. "Cape Coral Asks Court to Forbid Name to Rival," *News-Press* (FL), 24 August 1962.

57. "Cable System Sought for Cape Coral," *News-Press* (FL), 9 January 1964; Bernard, *Lies That Came True*, 173–75.

58. "Cape Coral Officials Say Its Buyers Are Satisfied," *News-Press* (FL), 24 December 1965; Ralph Weiskittel, "Developers Turning to 'Inner' Florida," *Cincinnati Enquirer*, 25 April 1971.

59. Paul D. Martin, "Port St. Lucie Turns 30," *Palm Beach Post*, 21 April 1991.

60. Clayton McMillon and Steve Anderson, *Sandstone Aquifer at Lehigh Acres Maximum Developable Limits* (West Palm Beach, FL: South Florida Water Management District, 2015), 2.

61. Eddie Pertuit, "Scenery Lush, Water Stinks along the Caloosahatchee," *News-Press* (FL), 15 November 1966.

62. "Pollution's Forces Press Attack by Air and Water," *Florida Today*, 28 June 1970.

63. "Expansion Program Told for Utilities Subsidiary," *Palm Beach Post*, 27 September 1962; "General Development Utilities Grow," *Orlando Sentinel*, 16 April 1961.

64. Kris Hundley, "A History Lesson in Governing Growth," *St. Petersburg Times*, 9 August 2009.

65. Mark Harper, "Deltona's Goal: Defend against Costly Septic Tank Rules," *Daytona Beach News-Journal*, 1 February 2015.

66. Jeannine Gage and Mark Harper, "How Did This Happen?—Deltona Flood Problems Run Deep," *Daytona Beach News-Journal*, 4 May 2003.

67. William Juneau, "Probers Hear Lot Sale Woes," *Chicago Tribune*, 5 April 1972.

68. U.S. Congress, Senate, Special Committee on Aging, *Interstate Mail Order Land Sales: Hearings before the Subcommittee on Frauds and Misrepresentations Affecting the Elderly*, 88th Cong., 2nd sess., 1964, 92.

69. Elizabeth Whitney, "Florida Land Sales Big Overseas," *St. Petersburg Times*, 6 January 1971.

70. "Mackles Go Long Way," *Orlando Sentinel*, 18 October 1964.

71. Sid Porter, "The Mackles," *Florida Magazine*, 20 March 1966, 8-F.

72. Frank Mackle III, "The Horses and the Farm," The Mackle Company, accessed 1 June 2017, http://www.themacklecompany.com/femjrstorypublic/personalstories/horses_and_the_farm.htm.

## Chapter Three

1. "Claude Kirk, Governor," *Tallahassee Democrat*, 8 November 1966.

2. Martin Dyckman, "Before Donald Trump, There Was Florida's Claude Kirk," *Floridapolitics.com*, 21 March 2017, http://floridapolitics.com/archives/234300-martin-dyckman-donald-trump-floridas-claude-kirk.

3. "Florida. I, Claudius," *Time*, 15 December 1967, 152.

4. Lucy Morgan, "Gov. Claude Kirk Set Bar for Florida Nuptial Drama in 1967," *St. Petersburg Times*, 11 December 2008.

5. "Colorful, Flamboyant Kirk Makes a Show of Politics," *Tallahassee Democrat*, 12 April 1970.

6. "Sen. Pope Favors Investigation of Exclusive 'Governor's Club," *Fort Lauderdale News*, 31 July 1968; "Kirk Honeymoon Paid for by FDC," *Tallahassee Democrat*, 7 December 1967.

7. "'Aquanut' Kirk Puts Claim Flag in Davey Jones Locker," *Palm Beach Post*, 26 October 1967.

8. Clarence Jones, "Kirk a High Flier as Business Man," *Tallahassee Democrat*, 15 December 1969; Dyckman, "Before Donald Trump, There Was Florida's Claude Kirk."

9. "Kirk Fumes at Cycler's 'Medal,'" *Orlando Sentinel*, 23 November 1967; "Colorful, Flamboyant Kirk Makes a Show of Politics"; "Fl. Governor to Aid Young Vendor," *Town Talk* (LA), 22 October 1967.

10. Barry Horenbein, interview by David E. Dodrill, Lee County Oral History Collection, University of Florida Digital Collections, 23 January 1990, 3.

11. "Bryant Names Five to Land Sales Board," *Tallahassee Democrat*, 20 July 1963.

12. "Land Sales Board," *News-Press* (FL), 30 September 1966.

13. Morton C. Paulson, *The Great Land Hustle* (Chicago: Henry Regnery, 1972), 166.

14. Lee Winfrey, "Big Names Sell Florida Swamp Lots," *Akron Beacon Journal*, 20 May 1964. See also David E. Dodrill, *Selling the Dream: The Gulf American Corporation and the Building of Cape Coral, Florida* (Tuscaloosa: University of Alabama Press, 1993), 167.

15. Juanita Greene, "Can Counties Cope with Sales Giant?" *Miami Herald*, 20 January 1966.

16. Larry Birger, "Faith for Sale: $800 an Acre," *Miami News*, 27 June 1965.

17. Birger.

18. Leonora LaPeter, "Off Road, Off Limits," *St. Petersburg Times*, 5 June 2005.

19. Jack Snyder, "Land Sales Company Pushes Unseen Lots," *Orlando Sentinel*, 24 September 1967.

20. Trevor Armbrister, "Land Frauds," *Saturday Evening Post*, 27 April 1963, 21.

21. George Eberl, "Florida: The Swamp and the Swamping," *Stars and Stripes*, 13 November 1968.

22. Susan Orlean, *The Orchid Thief* (New York: Ballantine, 1998), 130.

23. Frederic Sherman, "Selling the Swamp to the Suckers," *Detroit Free Press*, 4 April 1974.

24. U.S. Congress, Senate, *Interstate Land Sales Full Disclosure Act: Hearings before a Subcommittee of the Committee on Banking and Currency*, 89th Cong., 2nd sess., 1966, 22, 38–39; Michael Griffin, "No Place to Build a Home," *Orlando Sentinel*, 21 July 1992.

25. Phillip Meyer, "Rosen Cops Out as Witness at Mail-Order Land Probe," *Tallahassee Democrat*, 19 August 1966.

26. U.S. Congress, Senate, *Interstate Land Sales Full Disclosure Act*, 207–8.

27. Kenneth G. Slocum, "Furor in Florida: Land Sales Practices of Gulf American Corp. Touch Off Controversy," *Wall Street Journal*, 28 June 1967.

28. Dodrill, *Selling the Dream*, 106–7; Luther Evans, "Thomas Group Due Franchise," *Miami Herald*, 5 August 1965.

29. Eileen Bernard, *Lies That Came True: Tall Tales and Hard Sales and the Making of Cape Coral, Fla.* (Fort Myers: Press Printing, 1991), 224.

30. Slocum, "Furor in Florida."

31. "Land Sales Rules Backed by Mackle," *News-Press* (FL), 12 March 1967; "Sales of Flooded Land Are Banned," *News-Press* (FL), 7 September 1968; Haines Colbert, "One for All in Building an Empire," *Miami News*, 14 February 1968.

32. "Consumer Faces Rise in Fraudulent Sales," *Asbury Park Evening Press*, 15 September 1972.

33. Nathaniel P. Reed, interview with author, 1 July 2017.

34. Leonard Rich, interview by David E. Dodrill, Lee County Oral History Collection, University of Florida Digital Collections, 27 January 1988, 21.

35. Bernard, *Lies That Came True*, 50–52.

36. Geoffrey Birt, "Light Turnout of Electorate Is Seen," *Palm Beach Post*, 26 May 1964.

37. Edmund F. Kallina Jr., *Claude Kirk and the Politics of Confrontation* (Gainesville: University Press of Florida, 1993), 28; "Burns Fires Sharp Blast at Opponent," *Orlando Sentinel*, 6 May 1966; Rex Newman, "Kirk Continues Attack on High," *Pensacola News*, 25 October 1966.

38. Reed, interview with author.

39. Sylvia Porter, "Beware Mail-Order Land Gyps," *Asheville Citizen Times*, 20 August 1966.

40. Slocum, "Furor in Florida."

41. "Man Held in Kin's Death," *News-Press* (FL), 17 December 1966.

42. "Collier Grand Jury Impaneled," *News-Press* (FL), 13 December 1966.

43. Office of State Attorney David Aronberg, "Grand Jury Section," accessed 9 November 2020, http://www.sa15.state.fl.us/stateattorney/ouroffice/divisions/indexHGJ.htm.

44. Fred Winter, "Equalization Board Gets, Denies Four Tax Protests," *News-Press* (FL), 9 August 1966.

45. Vince Conboy, *Exposé: Florida's Billion Dollar Land Fraud* (Naples, FL: self-pub., 1972), 53.

46. Conboy, 50–51.

47. Conboy, 74.

48. "Land Sales Board Asks More Power," *Miami News*, 11 January 1967.

49. Kallina, *Claude Kirk and the Politics of Confrontation*, 12.

50. Malcolm B. Johnson, "Porkchop Gang's Origin and Nature," *Tallahassee Democrat*, 15 November 1959.

51. Seth A. Weitz, "Bourbon, Pork Chops, and Red Peppers: Political Immorality in Florida, 1945–1968" (PhD dissertation, Florida State University, 2007), vi.

52. John C. Gerard, "You Put 'Go' in 'Gravy Train,'" *Fort Lauderdale News*, 15 February 1961.

53. Ray Wieland, "What Reapportionment Will Mean," *Fort Lauderdale News*, 29 July 1962.

54. Malcolm B. Johnson, "Much for Puzzlement in Governor Campaign," *Tallahassee Democrat*, 30 October 1966.

55. Jack Roberts, "Good Old 'Claude, Baby' Kirk," *Miami News*, 26 January 1978; Charles Hesser and Milt Sosin, "Kirk Says High Fails to Fight City Crime," *Miami News*, 16 September 1966.

56. Slocum, "Furor in Florida."

57. Slocum.

58. Florida State Board of Conservation, Division of Geology, *Oil and Gas Activities in Florida, 1966*, Information Circular No. 55, by Clarence Babcock (Tallahassee: State Board of Conservation, 1968), 4.

59. "Trading Resumes in Gulf American," *Fort Lauderdale News*, 1 July 1967.

60. Ted Turpin, "Williams to See 'Gulf' Files," *Tucson Daily Citizen*, 7 July 1967.

61. Martin Waldron, "Billion-Dollar Installment Land Sales Business in Florida Faces Inquiry after Unethical Practices Charge," *New York Times*, 10 July 1967.

62. "Press Agent Denies Kirk Campaign Role," *Orlando Sentinel*, 9 May 1967.

63. "Kirk Schedules Trips to GOP Strategy Meet," *Pensacola Journal*, 26 September 1967.

64. "Presidential Bug Bites Governor," *Orlando Sentinel*, 10 November 1967.

65. "Marco Island Holds First Tony Lema Golf Tourney," *News-Press* (FL), 28 February 1967; "Nixon Buys Lot on Key in Dade," *Fort Lauderdale News*, 21 February 1967.

66. Kallina, *Claude Kirk and the Politics of Confrontation*, 23.

67. Dick Young, "Gulf American Space Center Tours Draw NASA Fire," *Orlando Sentinel*, 14 September 1967.

68. "Kirk Asks Development Chief to Quit," *Miami News*, 13 October 1967.

69. "Sen. Horne Hired by GAC to Handle Land Sales Case," *News-Press* (FL), 27 September 1967; "Horne Invites Official Inquiry," *Tallahassee Democrat*, 4 October 1967.

70. Rich, interview by Dodrill, 28.

71. Eddie Pertuit, "Rosen Suggests Gulf American Sale in Offing," *News-Press* (FL), 24 January 1968.

72. Clarence Jones, "Dale Was Victim of Florida System," *Tallahassee Democrat*, 19 December 1969.

73. Horenbein, interview by Dodrill, 5.

74. "Gulf American Is Charged with Fraud in Land Sales," *News-Press* (FL), 14 October 1967.

75. "Gulf American Says Guilty, Sales Barred for 30 Days," *News-Press* (FL), 11 November 1967.

76. Jim Hardee, "Hanky-Panky in Land Sales Over," *Tallahassee Democrat*, 15 November 1967.

77. "State Gets $1.4 Billion Lift from Gulf American Deals," *News-Press* (FL), 29 August 1967; "Size Helped GAC, Board Man Says," *News-Press* (FL), 20 November 1967; "Gulf American Suspension Lifting Refused," *Palm Beach Post*, 6 December 1967.

78. "Monitors Get $1,000 a Day," *Fort Lauderdale News*, 28 November 1967; "Gulf American Has Penalties Lifted," *News-Press* (FL), 3 February 1968; "Kirk Aides Named in Gulf Am Suit," *Fort Lauderdale News*, 27 January 1968.

79. "Land Sales Board Plans to Transfer Gulf American Case to Grand Juries," *Pensacola News Journal,* 8 February 1968; "Land Firm Files Petition," *Orlando Sentinel,* 30 March 1968; "Carl Bertoch Seeking State Defense Aid," *Tallahassee Democrat,* 6 February 1968; "Gulf Land Sales Ordered Halted in New Jersey," *Tallahassee Democrat,* 23 November 1967; "License Suspended," *The Progress* (PA), 8 December 1967; "Mack Quits Gulf American," *Miami News,* 2 May 1968; "Customer Sues to Keep Gulf American from Mails," *Fort Lauderdale News,* 5 March 1968; "Mack Quits Gulf American," *Miami News,* 2 May 1968.

80. "Land Firm Puts Loss at Million," *Fort Lauderdale News,* 14 April 1968; "Suspension Hurt, Says Gulf Am," *Miami News,* 23 October 1968.

81. "Land Sales Board Rejects Gulf American Offer," *Palm Beach Post,* 10 July 1968.

82. Dodrill, *Selling the Dream,* 214–15.

83. Rich Oppell, "Gulf American Gets Go-Ahead to Sell New Acreage," *Fort Lauderdale News,* 7 September 1968.

84. "Giant Airliner Due Here," *Tallahassee Democrat,* 11 November 1968.

85. Erik Calonius, "Deltona Says It's Back—Stronger than Ever," *Florida Trend* 23, no. 11 (March 1981): 63.

## Chapter Four

1. "Six Southwest Florida Residents among World's Billionaires," *Naples Daily News,* 22 March 2017.

2. *Lakeland Star-Telegram* op-ed reprinted in "Collier County an Asset," *Miami Daily Metropolis,* 13 April 1923.

3. Bruce D. Epperson, *Roads through the Everglades* (Jefferson, NC: MacFarland, 2016), 116.

4. Theresa Hamilton Proverbs, "We Built That: The Lost Fight for Florida's Cross-State Highway," *Journal of Planning History* 14, no. 4 (2015): 340; "County to Select Best Route for Road across Peninsula," *Miami News,* 28 June 1915.

5. G. E. Hosmer, "Barron G. Collier Tells the *Fort Myers Press* of Development Plans," *Fort Myers Press,* 7 May 1923.

6. Eric Collin, "Barron G. Collier, Empire Builder," *Tampa Tribune,* 10 May 1925.

7. Anthony Weitzel, "The Collier Story," *Naples Daily News,* 4 July 1976.

8. "Two Die in Dynamite Blast," *Fort Myers Press,* 24 August 1926.

9. "Collier County Fair," *Fort Myers Press,* 14 April 1928.

10. "Trail Caravan Wends Its Way to East Coast," *Tampa Daily Times,* 26 April 1928.

11. "Miami Planning Warm Welcome," *Miami Daily News,* 26 April 1928.

12. "Miami Land Bonds Are Auctioned Here," *New York Times,* 24 February 1928.

13. "Florida Is Facing Serious Reaction," *New York Times,* 3 January 1927.

14. "Collier Confidant State Development Can Not Be Checked, *Tampa Morning Tribune,* 11 March 1926.

15. "Tamiami Trail Beautification Plan Set Afoot," *Miami Daily News,* 27 April 1928.

16. Craig Woodward, "Joe Dickman: The Hermit of Kice Island?" *Coastal Breeze News,* 3 April 2014; Jeff Reynolds, "Progress Is Nudging Marco Island 'Hermit,'" *Miami News,* 15 February 1967.

17. "Marco Renamed San Marco by New Owners of Town," *Fort Myers Press*, 12 January 1926; Craig Woodward, "Key Marco," *Coastal Breeze News*, 22 April 2010.

18. "Edison Visit Collier City," *Fort Myers Press*, 12 May 1930.

19. Rufe Daughtrey, "Clam Industry, Dead 36 Years, May Start Up," *News-Press* (FL), 3 January 1955.

20. "Barron Collier's Creditors Accept Debt Payment Plan," *News-Press* (FL), 11 July 1933; "Tamiami Bus Line Resumes," *Tampa Daily Times*, 28 September 1935; "Gigantic Marsh Buggy Searches for Oil in Collier County Jungles and Swamps," *News-Press* (FL), 24 August 1937.

21. Bill Gentry, "Old 120: Marco Island's Hand-Operated Swinging Bridge," *Tampa Tribune-Times*, 19 October 1969.

22. Don Petit, "Collier City Is No More," *Miami News*, 29 September 1949; "Entire Village Is Moved on Marco Island," *News-Press* (FL), 16 February 1955.

23. Douglas Waitley, *The Last Paradise: The Building of Marco Island* (Coconut Grove, FL: Pickering Press, 1993), 23.

24. "Islands Will Have 4,000 Lots," *Fort Lauderdale News*, 22 September 1956.

25. Jane Wood, "They're Subdividing the Ten Thousand Islands Now," *Miami News*, 22 July 1956.

26. "Collier Holdings in Lee County Will Go on Sale," *News-Press* (FL), 11 January 1959.

27. Waitley, *The Last Paradise*, 7.

28. Michael Coleman, "The Birth of the Island," *Marco Island Sun Times*, 30 January–5 February 2014.

29. "Deeds for Sale of Marco Island Filed in Marco," *News-Press* (FL), 28 June 1964; "Marco Island Sale Sets Huge-Scale Development," *News-Press* (FL), 13 March 1964.

30. Waitley, *The Last Paradise*, 25–26.

31. "Man Made Lake Source of Marco Water Supply," *Naples Daily News*, 17 March 1970.

32. "Marco Island—Tremendous Undertaking," *Fort Lauderdale News*, 23 January 1965.

33. "Mackle Brothers Reveal Plans for Marco Island," *Hartford Courant*, 31 January 1965.

34. Juanita Greene, "Florida's Public Land: Going, Going . . ." *St. Petersburg Times*, 7 July 1963.

35. *Annual Report of the Chief of Engineers, U.S. Army, on Civil Works Activities*, vol. 1 (Washington, DC: U.S. Government Printing Office, 1968), vii.

36. C. M. Duke, "Army Engineers Keep Busy in Port," *New York Times*, 15 January 1961.

37. "Permit to Fill Bay Beach Asked," *Tampa Daily Times*, 14 December 1948.

38. "Developers Seeking Dredge, Fill Permit," *News-Press* (FL), 12 July 1964.

39. Waitley, *The Last Paradise*, 42.

40. John Woerpel, "What About Marco Island?" *Detroit Free-Press*, 17 April 1965.

41. "Marco Gets Crowds of Visitors," *News-Press* (FL), 1 February 1965; Waitley, *The Last Paradise*, 55.

42. "Marco Island Sales Reach $5.5 Million," *News-Press* (FL), 21 March 1965; "Marco Island Has Come a Long Way in 18 Months," *Fort Lauderdale News*, 10 September 1966; "Marco Bridge Rate Is Upped," *News-Press* (FL), 23 April 1968.

43. "Activity Lively at Marco Island as Club Debuts," *Fort Lauderdale News*, 12 February 1966; Jewell Dean, "Marco, Once Lonely, Undeveloped Island, Turned into Center of Spectacular Beauty," *News-Press* (FL), 18 September 1966.

44. Beatrice Hort Holmes, *History of Federal Water Resources Programs and Policies, 1961–70* (Washington, DC: U.S. Government Printing Office, 1979), 133.

45. Marquis Childs, "America's Heritage into Asphalt Desert?" *St. Petersburg Times*, 22 March 1967.

46. U.S. Congress, House of Representatives, Committee on Merchant Marine and Fisheries, *Hearings before the Subcommittee on Fisheries and Wildlife Conservation*, 90th Cong., 1st sess., 1967, 30.

47. Michael Grunwald, "Par for the Corps," *Washington Post*, 14 May 2006.

48. "Memorandum of Understanding between the Secretary of the Interior and the Secretary of the Army," *Code of Federal Regulations*, Title 33, Navigation and Navigable Waterways, Part 200 to End (Washington, DC: U.S. Government Printing Office, 1980), 710–11; John Noble Wilford, "Ecologists Challenging Army Corps of Engineers," *Arizona Daily Star*, 19 February 1972.

49. James Ryan, "Boca Ciega Bay Example of Bad 'Fills,'" *St. Petersburg Times*, 19 November 1968.

50. Bill Blalock, "Boca Ciega Bay Flaunted by Both Sides in Raging Fight over Man-Made Land," *Tampa Tribune*, 19 January 1960.

51. James Lewis, "Shrinking Bays Face New Inroads in Future," *St. Petersburg Times*, 21 March 1966.

52. "What Happens If We Do Nothing?" *St. Petersburg Times*, 10 May 1972; James Lewis, "Boca Ciega Bay One-Third Filled since Early 50's," *St. Petersburg Times*, 1 September 1964.

53. Stan Witwer, "Bayway Development: How It All Began," *St. Petersburg Times*, 17 September 1972.

54. "Unworkable Idea Becomes a Success," *St. Petersburg Times*, 29 April 1951; "Home Sales Again Heavy in Pinellas," *Tampa Tribune*, 19 December 1954.

55. "Slides Used in Protest against Fill," *St. Petersburg Times*, 29 January 1959.

56. "After All the Others, One More Fill Won't Hurt Boca Ciega Bay: Engineer," *St. Petersburg Times*, 28 March 1959.

57. Glenn J. MacGrady, "Florida's Sovereignty Submerged Lands: What Are They, Who Owns Them, and Where Is the Boundary?" *Florida State University Law Review* 1, no. 4 (1973): 607–8.

58. "Controversial Dredge-Fill Hearing Set," *St. Petersburg Times*, 6 July 1966.

59. James Ryan, "Boca Ciega Fill Plan Draws Heavy Protest," *St. Petersburg Times*, 30 November 1966.

60. James Lewis, "Boca Ciega Fill Plan Is Rejected," *St. Petersburg Times*, 16 March 1967.

61. Ben Funk, "'Black Fen' Now Dead River," *News-Press* (FL), 20 December 1966.

62. Lyle Van Bussom, "Desoto Legislator Says Pollution in Peace River 'Worst' Case in 50 Years," *Tampa Tribune*, 14 January 1960; "Peace River Fish Die," *Orlando Sentinel*, 15 March 1967; "The Ravaged River," *Tampa Tribune*, 20 March 1967.

63. Eddie Pertuit, "Scenery Lush, Water Stinks along Caloosahatchee," *News-Press* (FL), 15 November 1966; Ellen Stone, "Illness Prompts Ban on Area Shellfishing," *News-Press* (FL), 20 November 1973.

64. Larry Vickers, "Raw Sewage Dumped Daily into Caloosahatchee River," *Tampa Tribune*, 1 September 1967.

65. Betty Kohlman, "Dig Deep Enough in Florida, You'll Get Salt," *St. Petersburg Times*, 2 March 1969.

66. Sherri Blank, "Agency Seals Abandoned Wells to Stop Saltwater Intrusion," *News-Press* (FL), 14 June 1981; Tim Anderson, "Saltwater Intrusion Apathy May Be Costly," *News-Press* (FL), 3 October 1977.

67. Mark Stephens, "Water Crisis Threatens Future of South Florida," *News-Press* (FL), 9 August 1981.

68. Russell Clemmings, "Gulf Coast Has Water, but It's Not Cheap," *Sun-Sentinel*, 6 September 1983.

69. Clarke Ash, "Gov. Kirk's Smartest Move," *Miami News*, 22 April 1969.

70. Nathaniel Pryor Reed, *Travels on the Green Highway: An Environmentalist's Journey* (Hobe Sound, FL: Reed Publishing, 2016), 2–3.

71. "Submerged Land Sales Net State $2 Million," *Tampa Tribune*, 25 April 1966; "$250,000 Voted to Aid Fair Building Construction," *Tampa Tribune*, 19 July 1961; "State Defers Remodeling," *Tallahassee Democrat*, 6 September 1961.

72. Herschel Cribb, "Faircloth Wins 45-Day Halt in Sale of Submerged Land," *Tampa Tribune*, 1 February 1967; Martin Dyckman, "Conservation Triumph," *St. Petersburg Times*, 15 July 1967.

73. "Hodges Says Boards Lagging on Bulkhead Realignments," *Orlando Sentinel*, 11 June 1969.

74. Dick Young, "Bay-Fillers Get Greenlight," *Orlando Evening Star*, 26 April 1968.

75. Waitley, *The Last Paradise*, 95–102.

76. Harold Seneker, "Deltona in Deep," *Forbes*, 1 October 1976, 47.

77. Jerry Drake, "Marco Island: Paradise or Disaster? A U.S. Agency Decides," *Naples Daily News*, 31 August 1975; Frank Mackle III, "The Marco Permits," The Mackle Company, accessed 23 April 2018, http://www.themacklecompany.com/femjrstorypublic/17-deltona-marcopermits.htm; "Dredge, Fill Permit Issue for Marco Estates Section," *News-Press* (FL), 9 December 1969; Waitley, *The Last Paradise*, 102.

78. Drake, "Marco Island: Paradise or Disaster?"

79. Craig Pittman and Matthew Waite, *Paving Paradise: Florida's Vanishing Wetlands and the Failure of No Net Loss* (Gainesville: University Press of Florida, 2009), 49–50.

80. "Cabinet Approves Market-State Land Swap," *Tampa Tribune*, 13 October 1971.

81. Jim Steinberg, "A New Start for Deltona," *Miami News*, 9 December 1985.

82. Larry Birger, "Land Development Industry Is on a Rocky Road," *Miami News*, 3 January 1976.

83. "Diversification Helps Deltona Battle Slump," *Miami News*, 11 September 1976.

84. "'Back-to-Basics' Homes Boost Deltona Sales," *Miami News*, 24 March 1975; "Deltona Lost $6.5 Million," *Naples Daily News*, 27 March 1975; Seneker, "Deltona in Deep," 46–47; Drake, "Marco Island: Paradise or Disaster?"

85. Scott Derks, "Audubon Flails Deltona's Business Sense," *News-Press* (FL), 4 September 1975.

86. Derks, "Audubon Flails Deltona's Business Sense."

87. Rick Barry, "Unions Press for Marco Work," *Tampa Tribune*, 3 September 1975; Jerry Drake, "U.S., State Environment Groups Push for Denial of Marco Permits," *Naples Daily News*, 3 September 1975; Scott Derks, "Jobs Take Stand at Marco Hearing," *News-Press* (FL), 3 September 1975.

88. Jerry Drake, "Deltona Is Offered a Compromise," *Naples Daily News*, 4 September 1975; Derks, "Audubon Flails Deltona's Business Sense."

89. Jeffrey K. Stine, "Regulating Wetland in the 1970s," *Journal of Forest History* 27, no. 2 (April 1983): 70–72.

90. Jerry Drake, "Corps Denies Barfield Bay, Big Key Permits," *Naples Daily News*, 16 April 1976; Jim Steinberg, "Marco Island Almost Sank Deltona," *Miami News*, 9 December 1985; Erik Calonius, "Deltona Says It's Back—Stronger than Ever," *Florida Trend* 23, no. 11 (March 1981): 65.

## Chapter Five

1. Richard P. Dennis, "Entire Life Savings Invested in Notes," *News-Press* (FL), 26 January 1975.

2. This particular Homestate ad appeared in the Fort Myers *News-Press* on 2 May 1972. Between 1971 and 1974, the *News-Press* ran no fewer than 200 of these ads.

3. "Testimony of Mrs. Marian Moore before Florida House of Representatives Select Committee on Fraudulent Mortgage Sales, Fort Myers, FL, 27 August 1975," transcript, 86 (State Archives of Florida, Florida House Committee on Fraudulent Mortgage Sales, R-HREG, Container 79.80, Range No. 10015, Shelf C).

4. Dick Dennis, "Miami Bankruptcy Next Step for Continental Land Corp.," *News-Press* (FL), 15 November 1974.

5. Jonathan Kwitny, *The Fountain Pen Conspiracy* (New York: Knopf, 1973).

6. Kwitny, xi.

7. Michael Goldman, "Audit Slams Dickinson," *Tallahassee Democrat*, 16 October 1976; D. G. Lawrence, "State Endorses Security Manual," *Orlando Sentinel*, 30 September 1972.

8. "$12 Billion in Stocks Registered," *Naples Daily News*, 7 January 1973.

9. Fred Winter, "Sheriff Says International Frauds Reach into Collier," *News-Press* (FL), 13 July 1972.

10. "Nine More Arrested in Probe of Counterfeit Stock Ring," *Orlando Sentinel*, 17 April 1973; Denis Sneigr, "Spurious Bonds Sold to POWs, Broker Testifies," *Fort Lauderdale News*, 16 October 1974; "Oil Well Fraud Suspected," *Tallahassee Democrat*, 21 March 1974; Robert Schwartzman, "Investing in Churches Can Be Chancy," *Florida Today*, 18 May 1975.

11. "Family Fight with Inflation 'Staggering' Study Says," *Miami News*, 16 January 1974.

12. Thomas C. Rox, "Income Cut, Idleness Sour Retirement," *Tallahassee Democrat*, 7 July 1973.

13. "Southwest Florida Retirement Either a Dream or a Nightmare," *News-Press* (FL), 19 February 1974.

14. Clayton Reed, "The South Suncoast: A Growth Takeoff?" *St. Petersburg Times*, 10 June 1073.

15. "Southwest Florida Retirement Either a Dream or a Nightmare."

16. R. B. Plunkett Jr., "The New Poor: Dreamers Find Paradise Costs More than Social Security Pays," *Fort Lauderdale News*, 2 April 1978; Jim Dawkins, "Elderly Poverty Level May Hit 45 Per Cent," *Tampa Tribune*, 29 December 1971; Sandra Mathers, "Wintertime—and the Living's Not Easy for Retirees on Social Security," *Orlando Sentinel*, 14 November 1971.

17. Al Volker, "Retirement for Many Means Poverty," *Miami News*, 12 April 1971.

18. Dirk Hansen, "Your Savings: Insured, but Not Against Inflation," *Des Moines Register*, 19 May 1974.

19. "Remarks by Gerald A. Lewis, Comptroller of Florida, to the Summit Meeting on Securities, at Tallahassee, Florida, 15 May 1975," 3 (State Archives of Florida, Florida Division of Land Sales and Condominiums, Litigation Files, 1971–1977, Series 524, First Mortgage Files, Box 1, File Folder 17).

20. "Fiery Crosses Warm Klan Rally Spectators," *Orlando Sentinel*, 11 March 1957; Robert S. Bird, "Flaming Crosses Light Florida's Hate-Mongering," *Miami News*, 14 April 1957; Dennis Shawn Allen, "Race Relations and Integration in Citrus County, Florida" (master's thesis, Florida State University, 2013); Dan DeWitt, "A Past Full of Hate Unearthed," *St. Petersburg Times*, 5 July 2005.

21. Richard Morgan, "Beverly Hills Founder Friend of Elderly," *St. Petersburg Times*, 7 December 1969; Nell Woodcock, "Mackles Eyes Hernando for Huge Subdivision," *Tampa Tribune*, 28 October 1966; Elizabeth Whitney, "Huge Land Purchase Is All but Confirmed," *St. Petersburgh Times*, 5 June 1969; Betty Alexander, "Sugarmill: A $7 Million Bonus?" *St. Petersburg Times*, 12 July 1973.

22. "In the Matter of Miami Window Corporation," *Securities and Exchange Commission Decisions and Reports*, vol. 41, 1 June 1962 to 31 May 1964 (Washington, DC: U.S. Government Printing Office, 1967), 70–71.

23. "Citrus County Acreage Sold to Miami Firm," *Tampa Tribune*, 14 January 1970; Jim Tunstall, "Mortgage Suit Naming 3,200 Filed in Citrus," *Tampa Tribune*, 13 May 1978.

24. Denis Sneiger, "Securities Dealer Charged in Fraud," *Fort Lauderdale News*, 4 May 1974.

25. Richard P. Dennis, "Land Appraisals Totally Dishonest," *News-Press* (FL), 3 May 1975.

26. "Testimony of Mr. Richard P. Dennis before Florida House of Representatives Select Committee on Fraudulent Mortgage Sales, Fort Myers, FL, 27 August 1975," transcript, 30 (State Archives of Florida, Florida House Committee on Fraudulent Mortgage Sales, R-HREG, Container 79.80, Range No. 10015, Shelf C).

27. Sneiger, "Securities Dealer Charged in Fraud"; "Jail Convictions Won't Help Land-Fraud Victims," *News-Press* (FL), 9 May 1976.

28. Richard P. Dennis, "Miami Bankruptcy Next Step for Continental Land Corp.," *News-Press* (FL), 15 November 1974.

29. Richard P. Dennis, "Investors Appeal Court; Fear $3.5 Million Loss," *News-Press* (FL), 10 October 1975.

30. Elizabeth Whitney, "Arizona Land Schemes, along with Mob Ties, Spill Over into Florida," *St. Petersburg Times*, 1 September 1976.

31. Richard P. Dennis, "Final Creditor Meeting Called in Bankruptcy of Land Company," *News-Press* (FL), 27 June 1975; Richard P. Dennis, "Entire Life Savings Invested in Notes," *News-Press* (FL), 26 January 1975.

32. "Testimony of Mr. Richard P. Dennis before Florida House of Representatives Select Committee on Fraudulent Mortgage Sales, Fort Myers, FL, 27 August 1975," transcript, 14.

33. "Time to Clean House in Land Sales Agency," *News-Press* (FL), 6 February 1976.

34. Wayne Ezell, "Data Shows Laxity of Land Sales Division," *News-Press* (FL), 12 August 1975.

35. Ezell.

36. "Former Askew Partner Denies Any Preferential Treatment," *St. Petersburg Times*, 29 January 1976.

37. "'State Negligence' in Land Fraud Probe," *Tallahassee Democrat*, 11 June 1975.

38. "Fraudulent First Mortgages Issued to Southwest Florida Residents," Memorandum from Louis S. St. Laurent to Joseph P. D'Allesandro, n.p. (State Archives of Florida, Florida House Committee on Fraudulent Mortgage Sales, R-HREG, Container 79.80, Range No. 10016, Shelf C).

39. Robert Reilly, "Land Swindle Could Involve 30 to 40 Firms," *Arizona Republic*, 19 September 1976.

40. "Three in Mail Fraud Receive Suspended Sentences," *Fort Lauderdale News*, 15 June 1968; "12-Year-Old Mojave Co. Land Fraud Resurfaces," *Yuma Daily Sun* (AZ), 15 September 1975.

41. Richard P. Dennis, ""Life Savings Lost, Probably Forever," *News-Press* (FL), 26 January 1975.

42. Richard P. Dennis, "Complaints Continue to Mount in Alleged Mortgage Scheme," *News-Press* (FL), 8 February 1975.

43. Robert Bentley, "Dogged Pursuit Pays Off for Dennis," *News-Press* (FL), 25 May 1975.

44. Fitz McAden, "Land Fraud Halt Seen in 'Weeks,'" *Tampa Tribune*, 9 July 1975; "William Cotterell, "Land-Fraud Task Force Formed by State and Federal Officials," *St. Petersburg Times*, 16 May 1975.

45. "Land Sales Regulators Step Down," *Palm Beach Post*, 2 March 1976; "Dickinson, Land Sales Division Blamed in Fraud Ring," *Orlando Sentinel*, 11 July 1975.

46. "Arizona Prober Is New on Job," *News-Press* (FL), 1 September 1975.

47. Jerry Seper and John Winters, "Master Con Man Ned Warren Sr. Dies of Cancer in County Detention Ward," *Arizona Republic*, 11 October 1980; Don Bolles, "Promoter Warren Dances the Machiavellian Twist," *Arizona Republic*, 21 April 1967.

48. "Elite Opened Gates of City to Warren," *Arizona Republic*, 25 June 1978.

49. "The 'Godfather' of Land Fraud," *The Press Democrat* (CA), 24 March 1977.

50. Albert J. Sitter, "Four Key Tapes in Land Case Have Vanished," *Arizona Republic*, 14 June 1974; "Grand Jury Witness in Probe of Warren Slain Gang Style," *Arizona Republic*, 20 February 1975.

51. "Testimony of Slain Witness Told," *Casa Grande Dispatch* (AZ), 28 August 1975.

52. Michael McCloy, "Arizona Land Deals Rocked by Scandal," *Indianapolis News*, 14 May 1975.

53. Jack Swason, "Warren Convicted of Extortion," *Arizona Republic*, 26 September 1975; Frank Turco, "Warren Killed His Chance for Freedom and a New Life," *Arizona Republic*, 9 May 1978.

54. Richard P. Dennis, "New Billion-Dollar Land Fraud Uncovered in New Mexico," *Burlington Free Press* (VT), 7 December 1975.

55. Whitney, "Arizona Land Schemes, along with Mob Ties, Spill Over into Florida."

56. Email correspondence with Michael McHugh, 10 May 2018.

57. "Business Editor Resigns," *News-Press* (FL), 9 June 1976.

58. Charles Reid, "Land Fraud Bills Breeze through Legislature," *Pensacola News Journal*, 5 June 1976.

59. "Land Sales to Military in Europe $30 Million Annually," *Stars and Stripes*, 12 December 1972.

60. Larry Birger, "Florida Land Sales Just Aren't Booming Anymore, Sellers Say," *Miami News*, 11 June 1977.

61. Elizabeth Whitney, "Development of a Developer," *St. Petersburg Times*, 5 November 1972.

62. Charles Cates, "'City-in-the-Round' Plan Unveiled for Martin Board," *Palm Beach Post*, 4 February 1968.

63. Jack Alexander, *Rotonda: The Vision and the Reality* (Mineral, VA: Tabby House, 1995), 34.

64. Whitney, "Development of a Developer."

65. Elizabeth Whitney, "Allegations Face Rotonda West," *St. Petersburg Times*, 6 May 1972.

66. "Cavanagh Acquires PB Investment, Inc.," *Palm Beach Post*, 11 May 1969.

67. Charles Cates, "Rotonda Told FCD Opposes Any Drainage into Loxahatchee River," *Palm Beach Post*, 1 November 1968; "Rotonda Drainage Plans Opposed by Audubon Unit," *Palm Beach Post*, 6 December 1968; "Martin Again Opposes Rotonda Drainage Plans," *Palm Beach Post*, 11 December 1968; Whitney, "Development of a Developer."

68. "Vanderbilts Waste No Time Developing Charlotte Ranch," *Tampa Tribune*, 16 December 1951; Alexander, *Rotonda*, 23.

69. Lowell Brandle, "Vanderbilts Control Acreage," *St. Petersburg Times*, 5 September 1955; Milton Plumb, "Millionaire Vanderbilts Employ Practical Approach on 35,000-acre Florida Ranch," *Tampa Tribune*, 24 May 1953.

70. "Mackles Pay $4 million for Acreage," *Punta Gorda Herald* (FL), 1 January 1959.

71. "Land's Value Would Increase with Drainage," *Palm Beach Post*, 10 June 1973.

72. Jim Reeder, "Developers: Property Not Intended for Homesites," *Palm Beach Post*, 22 September 1976.

73. "Rotonda: Greener Grass on the Other Side of Florida," *Palm Beach Post*, 20 September 1970.

74. "We See Florida a Bit Differently," *St. Petersburg Times*, 10 April 1972.

75. Alexander, *Rotonda*, 58; "Come See the Great Shape We're In," *News-Press* (FL), 9 September 1970.

76. "City in Round Underway," *Tampa Tribune*, 17 October 1969.

77. Fred Girard, "Versatile Seagren Vaults into Big Money," *St. Petersburg Times*, 21 February 1973.

78. Robert Thomas, "Superstars Not So Super After All," *Orlando Sentinel Star*, 22 February 1973.

79. Tom McEwen, "The Morning After," *Tampa Tribune*, 20 February 1973.

80. Jack Craig, "Timing Boosts NBA over NHL," *Boston Globe*, 22 April 1973.

81. "Lady Superstars Join Men in Hunt for $$$," *Detroit News*, 27 January 1975; Pati David, "Robert Duvall Is Superest Superstar," *News-Press* (FL), 17 March 1975.

82. "Rotonda-Go-Round," *St. Petersburg Times*, 15 January 1979; "Bankruptcy Action Taken by Cavanagh," *Daily Herald-News* (FL), 18 February 1975; Pat Woodcock, "Rotonda Developers Named in Foreclosure Suit," *Sarasota Herald-Tribune*, 4 December 1974.

83. Dan Sewell, "Investors in Rotonda's Future Fear There Isn't Any," *St. Petersburg Times*, 22 February 1981.

84. Bob Schofield, "Lawmen Comb Swamps for Pilot of Pot Plane," *News-Press* (FL), 12 January 1974; Mike Fuery, "Pot Laden Plane Plows into Field," *News-Press* (FL), 12 January 1974.

85. Alexander, *Rotonda*, 115.

86. Jim Pratt, "Rotonda West Is Beset by Irate Owners' Suits," *News-Press* (FL), 17 February 1976; "FTC Accuses Cavanagh of Deceptive Sales Practices," *Miami News*, 26 September 1975.

87. "Land Firms Facing Tougher Sales Rules," *Florida Today*, 6 September 1973.

88. "Complaint in the Matter of Cavanagh Communities Corporation, et al.," in *Federal Trade Commission Decisions: Findings, Opinions, and Orders, January 1, 1979 to June 20, 1979* (Washington, DC: US Government Printing Office, 1980), 559–617.

89. John Cunniff, "Recession May Be Beneficial to Homebuilders' Future," *Fort Lauderdale News*, 10 February 1976; Tony Weitzel, "'Basic Homes' Prove Success for Deltona," *Naples Daily News*, 25 January 1976.

90. Frank Mackle III, "Rotonda—Cavanagh Communities," The Mackle Company, accessed 5 September 2018, http://www.themacklecompany.com/femjrstorypublic/otherstories/rotunda.htm.

91. Alexander, *Rotonda*, 59, 115; Elizabeth Whitney, "State: Developer Diverted $2.4 million," *St. Petersburg Times*, 21 September 1979; Jane Godown, "County Seeks Details about Rotonda West," *News-Press* (FL), 27 January 1975.

92. Alexander, *Rotonda*, 50–56.

93. Pat Woodcock, "Developers of Rotonda West Asked to Make Status Report," *Sarasota Herald-Tribune*, 11 December 1974.

94. Bob Schofield, "Charlotte Commissioner Blasts Rotonda Plans," *News-Press* (FL), 18 May 1977; Bob Schofield, "$109.7 Million Program Unveiled for Rotonda," *News-Press* (FL), 20 April 1977.

95. Elizabeth Whitney, "Cavanagh Is Selling Rotonda to Concentrate on Gambling," *St. Petersburg Times*, 15 June 1979; Pati Davis, "Rotonfa Deal Is Completed," *News-Press* (FL), 11 June 1980.

96. Alexander, *Rotonda*, 195–96.

## Chapter Six

1. David Olinger, "Hunting Club Takes Control of Polk Expanse," *St. Petersburg Times*, 9 January 1996; David Olinger, "Banned from Their Land," *St. Petersburg Times*, 14 January 1996.

2. Leonora LaPeter, "Off Road, Off Limits," *St. Petersburg Times*, 5 June 2005; Olinger, "Hunting Club Takes Control of Polk Expanse."

3. Dennis Covington, *Redneck Riviera: Armadillos, Outlaws, and the Demise of the American Dream* (New York: Counterpoint, 2004), 46.

4. Kimberly C. Moore, "Lynching, Klan, a Part of Polk History," *The Ledger* (FL), 5 May 2018; LaPeter, "Off Road, Off Limits."

5. There is a significant body of literature related to the impact of 1950s-era installment land sales developments in Florida. For a selection, see Timothy S. Chapin, Charles E. Connerly, and Harrison T. Higgins, eds., *Growth Management in Florida: Planning for Paradise* (London: Ashgate Press, 2007); Hubert B. Stroud, *The Promise of Paradise: Recreational and Retirement Communities in the United States since 1950* (Baltimore: Johns Hopkins University Press, 1995); Hubert B. Stroud and Bill M. Spikowski, "Planning in the Wake of Florida Land Scams," *Journal of Planning Education and Research* 19 (1999): 27–39; and Frank Schnidman and R. Lisle Baker, "Planning for Platted Lands: Land Use Remedies for Lot Sale Subdivisions," *Florida State University Law Review* 11, no. 3 (1983): 508–94.

6. Florida Department of Environmental Protection, *Florida Forever Five-Year Plan, 2017* (Tallahassee: DEP, 2017), 7, 72, https://fl.audubon.org/sites/default/files/may_2017_ff.pdf.

7. Ron Hurtibise, "Paper Paradise," *Daytona Beach News-Journal*, 9 January 2005.

8. Dara Kam, "Bombing Range Land under Fire," *Ocala Star-Banner*, 23 February 2000; Bill Bair, "Hunt Club Vows to Fight State," *The Ledger* (FL), 25 July 1999.

9. Hurtibise, "Paper Paradise."

10. Hurtibise.

11. "State Orders Land Seller to Give Full Refunds," *Tampa Tribune*, 5 June 2005.

12. Ludmilla Lelis, "Buyers of Swampland in Volusia County Hope to Get Their Money Back," *Orlando Sentinel*, 17 December 2011.

13. Rob Walsh, interview with author, 14 September 2018.

14. Eric Staats. "Battle between Development, Environment Has Been a Fully Engaged Struggle for Decades," *Naples Daily News*, 2 April 2012.

15. Greg Stanley, "Living in the Wild, Wild East." *Naples Daily News*, 17 April 2016.

16. Bill Belleville, "Lost City of the Everglades," *Sun Sentinel* (FL), 2 February 1992.

17. Lee Melsek, "Body Dump: The Mysterious 'Glades Holds Grisly Secrets," *News-Press* (FL), 4 December 1977.

18. "Collier Marijuana Growing Probe Continues," *News-Press* (FL), 16 August 1983; Denes Husty, "Deputies Seize Some 2,730 Marijuana Plants," *News-Press* (FL), 3 March 1988.

19. Denes Husty, "$175 Million in Cocaine Confiscated," *News-Press* (FL), 25 May 1984.

20. Mark Andersen, "Agents' Search for Marijuana Bales Unsuccessful," *News-Press* (FL), 19 December 1985.

21. Denes Husty, "Anti-Sandinistas Camped out in Collier," *News-Press* (FL), 15 February 1984; Denes Husty, "Collier Orders Bay of Pigs Veteran to Raise Military Camp," *News-Press* (FL), 25 February 1984.

22. David Fleshler, "Long-Abused Part of the "Glades Faces Difficult Restoration," *Sun-Sentinel* (FL), 29 September 2002.

23. Jeffrey Kahn, "Development Fuels Threat of Infernos," *Palm Beach Post*, 26 May 1985.

24. Kevin Lollar, "Restoration Project Is a Natural," *News-Press* (FL), 27 February 2001.

25. Betty Price, "Collier's Golden Gate: Can the Land Be Saved?" *News-Press* (FL), 23 January 1977.

26. U.S. Army Corps of Engineers, *Picayune Strand Restoration Project: Comprehensive Everglades Restoration Plan, Vol 1.* (Jacksonville, FL: U.S. Army Corps of Engineers, 2004), iv.

27. Mark Robichaux, "Miami Firms Profit from Land Markups," *Miami Herald*, 3 December 1987.

28. Kelly Toughill, "Stuck and Getting in Deeper," *News-Press* (FL), 16 December 1984.

29. Toughill; Leroy M. Melsek, "Swamp Dealer with Record Says Venture Is Clean," *News-Press* (FL), 13 April 1986.

30. "Land Agents Out of Boundaries," *News-Press* (FL), 20 March 1985.

31. Tracy Eaton, "Negotiator's in the Middle on Land Buys," *Miami Herald*, 3 January 1988.

32. "Boo-Bird Slugged by Condor Coach," *Austin Statesman* (TX), 7 February 1972.

33. Elizabeth Whitney, "Righting a Wrong or Throwing Good Money after Bad," *St. Petersburg Times*, 4 June 1989.

34. Joan Richardson, "Recruiter Barred from Land Suit," *Indianapolis Star*, 18 August 1968.

35. Daniel J. Foley, "Convicted Swindler Seeks Q-C Money," *Quad-City Times* (IA), 16 July 1978.

36. "Hunt Suit Claims Fraud, Extortion," *Victoria Advocate* (TX), 9 July 1980.

37. Richardson, "Recruiter Barred from Land Suit."

38. Whitney, "Righting a Wrong or Throwing Good Money after Bad."

39. Marilee Loboda Braue, "Fraud Investigator Asks Money for His Defense," *The Record* (NJ), 24 August 1994.

40. "GDC Foe Convicted of Tax Fraud," *St. Lucie News-Tribune* (FL), 20 July 1995; Trevor Armbrister, "You Don't Need This Guy's Help," *Reader's Digest* 139 (September 1991): 147.

41. Michael Cotes, "County Losing Patience with State's Wetland Buyout," *Naples Daily News*, 16 June 1996.

42. Jeffrey Hartdorn, interview with author, 4 October 2018.

43. Jim Leusner, "Daytona Lawyer to Admit Felony," *Orlando Sentinel* (FL), 23 September 1997; Belleville, "Lost City of the Everglades."

44. Jeff Klinkenberg, "When Good Faith Goes Bad," *Tampa Bay Times*, 11 October 1987.

45. "Gore Announced Funding of 31,000-Acre Land Buy," *Florida Today*, 18 June 1997.

46. Eric Staats, "State Hopes Its Willingness to Pay More for Land Will Speed Buyout," *Naples Daily News*, 21 May 1998.

47. Eric Staats, "Cabinet to Vote on Land Buyout in Rural Collier," *Naples Daily News*, 23 January 2001; Eric Staats, "Tribe Forced to Give up Land for Forced Restoration," *Naples Daily News*, 27 May 2005.

48. Eric Staats, "Miccosukee Indians Buy 800 Acres in Collier County," *Naples Daily News*, 25 December 1997; Caroline J. Keough, "Judge Orders Shutdown of Pump Polluting Glades," *Miami Herald*, 7 October 1999.

49. Billy Cypress, "Miccosukee Tribe: Listen to the Facts about Oil Well," *Miami Herald*, 5 July 1993.

50. Joni James, "Florida, Tribe at Odds on Land Issue," *Miami Herald*, 10 February 2003.

51. Roberto Santiago, "Landowner, State Fighting over His Home in Everglades," *Miami Herald*, 18 January 2005.

52. Eric Staats, "DEP Gives Southern Golden Gate Estates Man Extension to Move Out," *Naples Daily News*, 1 December 2005; Maya Bell, "Everglades Holdout Gives In," *Baltimore Sun*, 11 December 2005.

53. "New Model Homes Go on Display at Port Charlotte," *News-Press* (FL), 4 February 1959.

54. Walter Powers, "Sarasota Seeks More Long-Range Planning," *Tampa Tribune*, 25 January 1959.

55. Norma Jean Hill, "Constant Rains Disable Manatee Septic Tanks," *St. Petersburg Times*, 20 March 1959.

56. "Port Charlotte Area Incorporated," *News-Press* (FL), 17 June 1959.

57. Warren Fossdick, "North Port Charlotte Annexing Draws Notes of Apprehension," *St. Petersburg Times*, 30 January 1960; Walter Powers, "North Port Grows to City of 16,640 Acres," *Tampa Tribune*, 28 January 1960.

58. "Planning, Zoning Bills Are Forecast," *Tallahassee Democrat*, 16 June 1960; Dick Traynor, "Need for Planning Zone Laws Outlined by Research Firm," *Tallahassee Democrat*, 8 May 1961.

59. Roger Williams, "The Politics of Planning," *Florida Weekly*, 16 November 2017.

60. David E. Dodrill, *Selling the Dream: The Gulf American Corporation and the Building of Cape Coral, Florida* (Tuscaloosa: University of Alabama Press, 1993), 30; "County Planning Group Organized to Aid Expansion," *News-Press* (FL), 24 March 1959.

61. Pete Packett, "Planning Board Takes Up Study of Legislation," *News-Press* (FL), 1 April 1959.

62. Andrew Conte, "Community's Tombstone Could Read: GDC Buried Us in Residential Lots," *Indian River Press Journal* (FL), 5 February 1998.

63. Paul Reyes, *Exiles in Eden: Life among the Ruins of Florida's Great Recession* (New York: Henry Holt, 2010), 215.

64. Victor Hull, "Plats Pave Way for Problems," *Sarasota Herald-Tribune*, 7 July 1996; Florida Legislative Committee on Intergovernmental Relations, *Platted Lands* (Tallahassee: Florida LCIR, 2004), 4.

65. Spikowski Planning Associates, *Lehigh Acres Commercial Land Use Study*, report prepared for the Lee County Community Redevelopment Agency, May 1996, http://www.spikowski.com/documents-LehighAcres/LehighAcresCommercialLandUseStudy.pdf.

66. Pete Skiba, "Commercial Growth Incentives Possible," *News-Press* (FL), 2 October 2005.

67. Drew Winchester, "Past Promises, Current Challenges," *Cape Coral Daily Breeze*, 15 October 2011; Sarah Jarvis, "Housing, of All Types, Is on Its Way to Area Near Mariner High School," *News-Press* (FL), 22 January 2018.

68. Max Forgey, interview with author, 8 November 2018.

69. Hull, "Plats Pave Way for Problems."

70. Mary Jo Tierney, "Blaze Too Big for Volunteer Fire Fighters," *Miami Herald*, 20 May 1985.

71. Robin Benedick, "Palm Coast Families Rebuild amid Rekindled Fear of Fire," *Orlando Sentinel*, 17 May 1986.

72. Marcia Mattson, "Despite Rules Enacted after 1985's Fires, Flagler County Found Itself in Flames Again," *Florida Times-Union*, 27 July 1998; Dinah Voyles Pulver, "Worst of Nature, Best of People," *Daytona Beach News-Journal*, 10 June 2018.

73. Maya Bell and Christine Shenot, "Florida Is Not Ready to Face New Fires," *Orlando Sentinel*, 26 July 1998.

74. Charles Runnells, "Neighborhoods without Hydrants Left Vulnerable," *News-Press* (FL), 13 April 2006.

75. Aaron Hale, "Lack of Fire Hydrants Hampering Firefighters in Some Areas of Collier," *Naples Daily News*, 19 September 2010.

76. "Traveling Asphalt Plant," *Fort Lauderdale News*, 10 June 1961.

77. Pat Harbolt, "Commissioners Allocate Money for Street Paving," *Tallahassee Democrat*, 31 August 1975.

78. Teresa Lane, "Port St. Lucie Backs Builder on Road Issue," *Palm Beach Post*, 20 April 1989.

79. Tom Frame, interview with author, 14 September 2018.

80. Teresa Lane, "City Celebrates Rise from Population 0," *Palm Beach Post*, 28 April 1996.

81. "Sarasota County's Roads Inadequate," *Tampa Bay Times*, 24 July 1957; Don Showerman, "Soil Cement Paving Decision Delayed," *Fort Pierce News-Tribune*, 7 January 1959.

82. Don Showerman, "County Shies Away from Okaying Soil Cement Roads," *Fort Pierce News-Tribune*, 31 December 1958; Don Showerman, "No Soil-Cement Roads—for the Present," *Fort Pierce News-Tribune*, 13 January 1959.

83. John Gordon, "Charlotte Accepts Roads in Platted Subdivisions," *Tampa Tribune*, 26 November 1964; "County to Maintain Streets in Port Charlotte," *News-Press* (FL), 15 December 1965.

84. Bill Sloat, "Bad Roads Need Facelift, Funds," *News-Press* (FL), 15 April 1976; "County to Maintain Roads in Port Charlotte," *News-Press* (FL), 15 December 1965.

85. Pete Johnson, "The County Will Fix Port Charlotte Roads," *News-Press* (FL), 21 September 1977; Tom Brew, "Commission Tackles Road Repairs," *Daily Herald-News* (FL), 21 September 1977.

86. Gordon Russell, "North Porters Receive an Apology," *Sarasota Herald-Tribune*, 30 November 1996; Jose Luis Jimenez, "Two Bodies Discovered in Woods," *Sarasota Herald-Tribune*, 18 April 1996.

87. Scott Derks, "Roads to Nowhere: Street Maze Awaits Homes," *News-Press* (FL), 27 April 1975.

88. "Capsules," *Florida Today*, 28 July 1979; "Deltona Sells Award-Winning Hotel," *News-Press* (FL), 20 February 1979; Jim Steinberg, "A New Start for Deltona," *Miami News*, 9 December 1985; Jim Steinberg, "Deltona Is Seeking New Chief Executive," *Miami News*, 5 February 1986.

89. Mike Clary, "Florida's Old Enemy—Land Fraud—Makes an Appearance," *Los Angeles Times*, 6 May 1990.

90. Amy Driscoll, "Fannie Mae, Freddie Mac Won't Buy GDC Mortgages," *Palm Beach Post*, 11 February 1990.

91. Jack Nease, "Developer Appears to Whitewash Filing," *Sun-Sentinel* (FL), 22 March 1990.

92. Carl Hiaasen, "Buying a Piece of Florida? See It Like a Native," *Miami Herald*, 16 February 1987.

93. Teresa Lane, "Jury Investigates General Development Appraisals," *Palm Beach Post*, 29 April 1988; "GDC under Siege," *Palm Beach Post*, 2 February 1990.

94. Alan Goldstein, "Developer May Be Indicted," *St. Petersburg Times*, 1 February 1990.

95. Amy Driscoll, "Homeowners' Lawsuits Contest GDC Appraisals," *Palm Beach Post*, 18 February 1990.

96. Joyce Harris, "Disagreement Sinks GDC-Deltona-Merger," *Florida Today*, 2 September 1988; "GDC Out of Land Resources Deal," *Florida Today*, 27 April 1989; Joyce Harris, "Developer under Dual Fire," *Florida Today*, 18 December 1989.

97. Michael Huber, "Florida Probes Company's Overseas Operations," *Miami Herald*, 5 March 1990.

98. Michael Huber, "GDC Says Officers May Face Fraud Charges" *Miami Herald*, 1 February 1990.

99. "General Development Debt Downgraded," *Palm Beach Post*, 13 March 1990; Michael Huber, "Investors Flee GDC Stocks," *Miami Herald*, 13 March 1990.

100. "Trash, Taxes Make List of GDC Topics," *Orlando Sentinel*, 28 April 1990; Paul D. Martin, "Port St. Lucie Seeks New Deal with GDC," *Palm Beach Post*, 5 June 1991; Myke Millano, "County Rejects GDC Stock," *News-Press* (FL), 20 November 1991; "City Stuck with Bill for Road Repair," *Orlando Sentinel*, 11 April 1990.

101. Susan Jacobson, "GDC Communities Stuck with Road Repair," *Orlando Sentinel*, 26 August 1990.

102. Joyce Harris, "GDC Offers Creditors Its Inactive Stock," *Florida Today*, 26 August 1991.

103. Teresa Lane, "Cities See Utilities Issue as a Struggle for Control," *Palm Beach Post*, 12 February 1989.

104. "GDC Considers Bankruptcy," *News-Press* (FL), 3 April 1990.

105. Scott Shifrel, "Glades County Inmates Left Alone 8 Hours a Night," *Palm Beach Post*, 2 May 1990.

106. Craig S. Palosky, "Tax Certificates Generate Little for North Port," *Tampa Tribune*, 29 September 1991.

107. Victoria Reid, "GDC Sentencing Hits Home," *Florida Today*, 6 January 1993.

## Chapter Seven

1. Dick Greco, interview with author, 7 January 2014.

2. Cynthia Mayer, "Silver and Old—Nation's Oldest County Is Heaven for Retirees, Dullsville for Young," *St. Petersburg Times*, 10 August 1987.

3. Robin Pollack, "Commission Repeals Tax for Parks," *News-Press* (FL), 5 August 1987.

4. Margo Harakas, "Florida's Elderly," *Palm Beach Post*, 3 June 1982.

5. Ken Zapinski, "Seniors Redefining Scope of Politics," *Hernando Times*, 21 October 1990.

6. Bill Stevens, "'Ageism' Strikes Deep with Vote to Ban Younger Residents," *St. Petersburg Times*, 19 April 1988; Diane Stallings, "Subdivision Considering Age Limits, Approval Plan," *St. Petersburg Times*, 3 April 1988.

7. Sid Kirchheimer, "South Florida's Generation Gap," *Sun-Sentinel* (FL), 1 March 1987. Quote by Tamarac mother from Grant Hanessian, "Tamarac: Generations Clash in the City That Wasn't Meant to Be," *Fort Lauderdale News*, 4 April 1977.

8. Deborah Bacon, "Elderly May Consolidate Growing Political Clout," *Tampa Tribune*, 1 August 1987.

9. "Survey: Older Voters Fed Up with Politics," *Sun-Sentinel* (FL), 8 July 1992.

10. "UF Study Says Older Voters Tend to Oppose School Bonds," *St. Petersburg Times*, 10 January 1991.

11. Ad Hudler, "Deputies Watch over Calmer Crowd," *News-Press* (FL), 5 August 1987; Deborah Bacon, "Elderly May Consolidate Growing Political Clout."

12. Tom Fiedler, "Republican Invasion Changes Political Face of Central Florida," *Miami Herald*, 29 October 1986.

13. Mark Journey, "Pinellas County Leading State in Number of Registered Republicans," *St. Petersburg Times*, 14 October 1988; George Wilkens, "GOP Gaining on Democrats," *Tampa Tribune*, 24 January 1992.

14. Ad Hudler, "Punching Out," *News-Press* (FL), 4 September 1988; "Charlotte County Results," *St. Petersburg Times*, 7 November 1956; "Democrats Leads Here 8 to 1," *News-Press* (FL), 12 October 1956.

15. "Port Charlotte GOP Protests Tax Increase," *News-Press* (FL), 25 April 1961; "One-Man Crusade Swings Port Charlotte Precinct over to GOP," *News-Press* (FL), 14 November 1960.

16. Roger Mullen, "Port St. Lucie Now Younger than Ever," *Palm Beach Post*, 5 May 1985.

17. Rock Tonyan, "Commuters May Hold Key to City of Deltona," *Orlando Sentinel*, 30 August 1990; Rick Tonyan, "Delayed Deltona I-4 Exit Back on 1991 Schedule," *Orlando Sentinel*, 2 March 1989.

18. "Low Land Prices and Natural Beauty Abound," *Orlando Sentinel*, 23 September 1990.

19. Alan Snel, "Port St. Lucie Struggles to Keep Up with Young Residents," *Palm Beach Post*, 28 June 1992.

20. "Loud 'Hokey Pokey' Gets Florida Bar Busted, and That's What It's All About," *Akron Beacon Journal* (OH), 5 March 1992; Teresa Lane, "Noisy Guests Could Cost Hotel $1,000," *Palm Beach Post*, 4 January 1988; Stephanie L. Jackson, "Port St. Lucie Noise Ordinance May Need Some Polishing," *Palm Beach Post*, 2 May 1992.

21. Teresa Lane, "Residents, Planners Oppose Deaf Center Site," *Palm Beach Post*, 6 October 1998; "Polarize or Compromise," *Palm Beach Post*, 30 November 1992.

22. "Skateboard Ramp OK in Yard, Judge Rules," *Palm Beach Post*, 4 March 1986; Jim Reeder, "St. Lucie, Stadium Developer Reach Pact on Tourist Revenue," *Palm Beach Post*, 31 December 1985.

23. Andrew Conte, "Big Apple Is Rotten to Port St. Lucie," *Indian River Press Journal* (FL), 14 November 1998.

24. Michael Dunn, "Jump Start," *Tampa Tribune*, 2 November 1985.

25. Sam Roberts, "Seeing the Boys of Summer in Spring," *New York Times*, 3 March 1991.

26. Paul D. Martin, "Port St. Lucie Turns 30," *Palm Beach Post*, 21 April 1991.

27. Paul D. Martin, "It's a Boy! City Welcomes First Newborn," *Palm Beach Post*, 9 January 1992. In 2011, *Sarasota Herald-Tribune* reporter Thomas Becnel made the same "There is no there there" reference when describing Port St. Lucie's sister city North Port. Thomas Becnel, "North Port: Underdog City That Gets No Respect," *Sarasota Herald-Tribune*, 15 August 2011.

28. Teresa Lane, "Port St. Lucie's First Cemetery to Open Soon," *Palm Beach Post*, 4 March 2004.

29. Leanora Minai, "Little Red Schoolhouse May Close," *St. Petersburg Times*, 26 May 1995; Tara Beer, "At the Polls, Comments Favor Library," *St. Petersburg Times*, 9 November 1994.

30. Doug Stanley, "Brooksville, Spring Hill Are Opposites," *Tampa Tribune*, 27 August 1990.

31. Stanley.

32. "Real Estate by the Numbers," *St. Petersburg Times*, 28 April 1997.

33. Saundra Amrhein, "Hispanics, Asians Put New Face on County," *St. Petersburg Times*, 23 May 1991.

34. Phyllis Lipinski, "Spring Hill Teen Club Drops Deal with Bubba," *St. Petersburg Times*, 19 July 1995.

35. Jeffrey S. Solochek, "Bubba, Business Partner Say County Bullied Club," *St. Petersburg Times*, 9 December 2001; Jennifer Farrell, "Planet Bubba Owners Vow Music Will Boom Louder," *St. Petersburg Times*, 19 April 2000.

36. Dan DeWitt, "Report: Parkway Numbers Are Low," *St. Petersburg Times*, 31 October 2002.

37. Michael Kruse, "Driving Growth," *St. Petersburg Times*, 5 February 2006.

38. Dan DeWitt, "Hernando County: Homes 'R Us," *St. Petersburg Times*, 24 May 2002.

39. Ryan Hiraki, "City Considers Imposing New Fee for Pricey Homes," *News-Press* (FL), 8 November 2005.

40. "Minority Buyers Lift Home Sales," *News-Press* (FL), 29 January 1998.

41. Bob Rathgeber, "First Home Builders Snaps up 714 Lots in Lehigh Acres," *News-Press* (FL), 22 January 1998.

42. Pamela Reeves, "Proposal Could Help High Income, Marginal Buyers," *News Journal* (DE), 14 March 1998.

43. Elisa Rocha, "Grants Open Doors to Homes," *Sacramento Bee*, 8 April 1996.

44. "More Down Payments Free," *News-Press* (FL), 5 January 2003.

45. Loretta Klab, "HUD Error Scares Area Housing Program," *Sacramento Bee*, 20 June 1997; U.S. Congress, House of Representatives, Subcommittee on Housing and Community Opportunity of the Committee on Financial Services, *Homeowner Down Payment Assistance Programs and Related Issues*, 110th Congress, 1st sess., 2007, 102.

46. U.S. Department of Housing and Urban Development, Office of the Inspector General, *Final Report of Nationwide Audit—Down Payment Assistance Programs*, 2000-SE-121-0001 (Washington, DC: HUD, 2000), 2–4, https://archives.hud.gov/offices/oig/reports/internal/ig000001.pdf.

47. Michael Corkery, "Housing Ministry: Scrutiny of Down-Payment Gifts Threatens Charitable Movement," *Wall Street Journal*, 5 July 2006.

48. Joan D. LaGuardia, "Filling Local Housing Gap: Program Would Allow Buyers Low Closing Costs," *News-Press* (FL), 12 June 1998.

49. Dick Hogan, "Housing Hits New Heights for '98," *News-Press* (FL), 21 January 1999.

50. Bob Rathgeber, "Home Sales Set Record in June," *News-Press* (FL), 31 July 1998; Don Ruane, "Lehigh Corp. Dethroned as Site Sales King," *News-Press* (FL), 7 September 1998; "State First Achieved by Builder," *News-Press* (FL), 9 August 1999.

51. Dick Hogan, "First Home Tops List of Busy Builders," *News-Press* (FL), 2 March 2003; Dick Hogan, "National Builders Dominate Lee," *News-Press* (FL), 14 August 2005.

52. "First Home Builders Honored on Local and National Level," *News-Press* (FL), 19 May 2005; Dick Hogan, "Program Helps Buyers Get Home with No Money Down," *News-Press* (FL), 22 November 2000.

53. "Home Sales: Number Down, Prices Up," *Charlotte Sun* (FL), 29 November 2005.

54. Joe Newman, "Developers' Legacy Costs Millions to Fix," *Orlando Sentinel*, 22 June 2003.

55. Laura Layden, "Their Favorite Year: Realtors and Developers Are Gushing over the Successes They've Seen in 2004," *Naples Daily News*, 26 December 2004.

56. Riddhi Trivedi-St. Clair, "More than 1,000 Seeking a Chance to Buy Home in Hawthorne Community," *The Banner* (FL), 18 June 2005; Robyn A. Friedman. "Going Like Hotcakes," *Sun Sentinel* (FL), 16 June 2002; Layden, "Their Favorite Year."

57. Kathryn Helmke, "Roofers Nailed by Supply Cracks," *Naples Daily News*, 13 August 2005. Gazzoli quote from Joe Crews, "Builders Taking Longer to Finish New Homes Force Buyers to Endure," *Daytona Beach News-Journal*, 28 December 2003.

58. Steven Ray Haberlin, "Building Blocks," *Ocala Star-Banner*, 4 December 2005; Robyn A. Friedman, "Buildup in Price," *Sun Sentinel* (FL), 22 September 2005. Hoyt quote from Riddhi Trivedi-St. Clair, "Construction Costs Keep on Climbing," *Naples Daily News*, 2 April 2006.

59. Michael Van Sickler, "Investors Drive Hot Housing Market," *St. Petersburg Times*, 17 July 2005; Jeff Ostrowski, "Speculators Cash In," *Palm Beach Post*, 31 July 2005.

60. Linda Rawls, "Housing Boom Makes Art of the 'Flip' Enticing," *Palm Beach Post*, 18 July 2004.

61. Gina Edwards, "Fast Bucks," *Naples Daily News*, 10 July 2005.

62. Ostrowski, "Speculators Cash In"; Melissa Followell, "Demand for Housing in Bradenton Fuels Idea for Class," *Bradenton Herald*, 27 July 2005. Connecticut man quote from Michael Kruse, "'Flippers' Play the Odds in Housing Boom," *St. Petersburg Times*, 12 July 2005.

63. Linda Rawls, "Used-Home Sales Soaring," *Palm Beach Post*, 26 August 2003.

64. Scott Blake, "Low End of Market 'Suffering,'" *Florida Today*, 25 May 2005; Riddhi Trivedi-St. Clair, "Labor Daze," *Naples Daily News*, 31 July 2005; Edwards, "Fast Bucks."

65. Riddhi Trivedi-St. Clair, "When It Comes to the Price of Housing in Lee County, the Numbers Tell the Story," *The Banner* (FL), 24 April 2004.

66. Dick Hogan, "Big Bucks Brace Lee Housing," *Florida Today*, 11 August 2005; Denes Husty III, "Fire Department Recruits Find It Hard to Buy Houses," *News-Press* (FL), 13 November 2005.

67. Paul Skiba, "US Home Changes Its Sale Strategy," *News-Press* (FL), 14 December 2005.

68. Elizabeth Wright, "Federal Charges Not All for Real Estate Agent," *Naples Daily News*, 24 June 2008. For a review of the First Home Lease Purchase Investment Opportunity scheme, see Alyssa Katz, *Our Lot: How Real Estate Came to Own Us* (New York: Bloomsbury, 2009), 163–67.

69. Kenneth R. Harney, "Rising Home Debts Aren't Crippling Borrowers—Yet," *St. Petersburg Times*, 26 March 2005; U.S. Congress, Senate, *The 2007 Joint Economic Report*, 110th Cong., 1st sess., 2007, S. Rep. 110–251, 25–26, https://www.jec.senate.gov/public/_cache/files/618bf038-5dad-4ff1-b7ad-e685c2b6bb77/2007annualreport.pdf.

70. Betty Parker, "Lehigh Paces Lee's Land Bonanza," *News-Press* (FL), 1 June 2005.

71. Scott Burns, "Florida Housing Market Forecast: Continued Sunny," *Dallas Morning News*, 21 March 2004; Michael Pollick, "Flipping Property," *Sarasota Herald-Tribune*, 12 July 2004; Michael Braga, "And the Condos Keep on Coming," *Sarasota Herald-Tribune*, 6 November 2005.

72. Ostrowski, "Speculators Cash In."

73. Sue Kirchhoff, "Home Prices Rise to 'Extremely Overvalued' Levels in 53 Cities," *USA Today*, 17 August 2005.

74. Linda Rawls, "Prices Hold as Homes Languish," *Palm Beach Post*, 28 June 2006.

75. Linda Rawls, "Home Sales Bust the Boom," *Palm Beach Post*, 26 July 2006.

76. Dick Hogan, "Lee Homes Prices, Sales Dive in Dec.," *News-Press* (FL), 26 January 2007.

77. Dick Hogan, "How Hovnanian Lost Big in Lee," *News-Press* (FL), November 2007.

78. Dick Hogan, "Dispute in Housing Deal Could Lead to Class-Action Lawsuit," *News-Press* (FL), 26 April 2007.

79. "Hovnanian Posts Another Loss," *News-Press* (FL), 7 September 2007. Sorsby quote from Dick Hogan, "Local Housing Sales Bleak," *News-Press* (FL), 26 June 2007.

80. Dick Hogan, "Bad Loans on Houses Are Being Negotiated," *News-Press* (FL), 24 September 2007; Dick Hogan, "Embattled Builder Cuts 200 Workers," *News-Press* (FL), 30 October 2007; Dick Hogan, "Builder Wrote off 1,345 Lee Homes," *News-Press* (FL), 12 March 2008; Dick Hogan, "First Home Sells Last Lots," *News-Press* (FL), 28 August 2009.

81. Eve Samples, "Levitt's Tumble Turns Enclave to 'Ghost Town,'" *Palm Beach Post*, 29 November 2007.

82. Mark Landler, "I.M.F. Puts Bank Losses from Global Financial Crisis at $4.1 Trillion," *New York Times*, 21 April 2009.

83. Christopher S. Rugaber, "Jobless Claims Off; Hiring Still Weak," *Philadelphia Inquirer*, 1 January 2010.

84. Jeff Harrington, "State Near Jobless Record," *St. Petersburg Times*, 23 January 2010.

85. "AP Stress Index Shows Twenty Most Stressed Counties in September," *Associated Press*, 2 November 2009.

86. Laura Layden, "Fort Myers-Cape Coral No. 1 in 2008 for Foreclosures," *Naples Daily News*, 15 January 2009; Janine Zeitlin, "Lee's Homeless Being Counted," *News-Press* (FL), 28 January 2010; Francesca Donlan, "A Crisis in Our Midst: Thousands Need Help," *News-Press* (FL), 10 May 2009.

87. Pat Gillespie, "Housing Woes Exacerbate the Despair of Divorce," *News-Press* (FL), 2 February 2009; Dan Warner, "Suicides Intensify in Crushing Times," *News-Press* (FL), 27 June 2009; Janine Zeitlin, "Child Abuse Up in Hard Times," *News-Press* (FL), 24 May 2009; Dan Ruane, "Unfinished Homes Start to Crumble," *News-Press* (FL), 23 April 2008.

88. Chris Umpierre, "Bankruptcies by Thousands Flood Courts," *News-Press* (FL), 20 December 2009; Janine Zeitlin, "Violence at Home," *News-Press* (FL), 27 December 2009; Dick Hogan, "What You Can Buy," *News-Press* (FL), 17 May 2009; Rachel Revehl, "Drug Deaths Rise Sharply," *News-Press* (FL), 6 July 2010; Jennifer Booth Reed, "Numbers of Uninsured Reach Epidemic Levels," *News-Press* (FL), 23 March 2009; Donlan, "A Crisis in Our Midst."

89. Donlan, "A Crisis in Our Midst."

90. Kris Hundley, "A History Lesson in Governing Growth," *St. Petersburg Times*, 9 August 2009.

91. Bob Rathgeber, "Housing Market Harbors Crime," *News-Press* (FL), 24 February 2009.

92. Evan Williams, "An Entrepreneur Finds a Booming Business: A/C Security," *Fort Myers Florida Weekly*, 17 February 2010; Mel Toadvine, "Heavy Steel Cages Protect AC Equipment," *Lehigh Acres Citizen*, 19 March 2010.

93. Peter Franceschina, "Residential Homes Fertile Ground for Pot Growers," *News-Press* (FL), 18 March 2001.

94. Franceschina. After a review of court records and interviews with law enforcement officials, in 2009 the *Orlando Sentinel* estimated that "up to 90 percent" of arrested

grow house suspects were immigrant Cubans. "Marijuana Goes Up in Smoke in Osceola," *Orlando Sentinel*, 23 December 2009.

95. Don Ruane, "Indoor Pot Growing an Issue in Lee," *News-Press* (FL), 20 April 2008.

96. "The Marijuana Wars," *Florida Today*, 12 February 2008.

97. Laura Layden, "Fort Myers-Cape Coral No. 1 in 2008 for Foreclosures," *Naples Daily News*, 15 January 2009.

98. Laura Layden, "Home, Condo Sale Prices Still Falling in Lee," *Naples Daily News*, 25 November 2008.

99. Roger Williams, "The New Price of Paradise," *Fort Myers Florida Weekly*, 13 January 2010; "Potential Home Buyers Float on By," *St. Petersburg Times*, 30 March 2009.

100. Dick Hogan, "Builders Pull 20 Permits," *News-Press* (FL), 5 January 2010.

101. Jacob Ogles, "Building Permits Crawl to Near Stop," *News-Press* (FL), 27 December 2007.

102. Ryan Lengerich, "Courts Tackle Housing Crisis," *News-Press* (FL), 1 December 2008.

103. Dick Hogan, "Push Is on to Clear Backlog of Foreclosure Cases in Lee," *News-Press* (FL), 5 December 2008; David Dayen, *Chain of Title: How Three Ordinary Americans Uncovered Wall Street's Great Foreclosure Fraud* (New York: The New Press, 2017), 257.

104. George Packer, "The Ponzi State," *New Yorker*, 9 and 16 February 2009, 85.

## Epilogue

1. Mike Fuery, "Ranch Has Water Supply Needed by County," *News-Press* (FL), 12 July 1973; Mark Stephens, "Cattle Stealing Rustles Up Big Money Here," *News-Press* (FL), 9 September 1979; Wes Smith, "Coalition Rallies behind Plan to Save Ranch," *Tampa Tribune*, 13 September 2004.

2. Frances D. Williams, "Wilderness Worker," *News-Press* (FL), 7 March 1993.

3. Pamela Smith Hayford, "Babcock Land Moves to Top of Acquisition List," *News-Press* (FL), 26 December 2001.

4. Jamie Manfuso, "Landowner Proposes New Town," *Sarasota Herald-Tribune*, 17 May 2002.

5. Jane Baty, "Lee, Charlotte Officials Talk Babcock," *North Port Sun*, 19 April 2002.

6. Greg Martin, "Groups Combine to Buy Babcock," *Charlotte Sun* (FL), 24 March 2004.

7. Chad Gillis, "Babcock Deal Inches Closer," *Charlotte Sun* (FL), 16 November 2004).

8. Greg Martin, "Campaign Envisioned to Buy Babcock," *Charlotte Sun* (FL), 27 May 2004.

9. Greg Martin, "DEP Explains Developer in Babcock Deal—Proposal Represented Brainstorming," *Charlotte Sun* (FL), 20 February 2005; "State's Back Door Dealing on Babcock Needs More Scrutiny," *Desoto Sun* (FL), 8 April 2005.

10. Jamie Page, "Charlotte Backs Babcock Plan," *News-Press* (FL), 15 December 2005.

11. John Dorschner, "New City in Florida to Run on Solar Power, *Miami Herald*, 10 April 2009.

12. John Haughey, "Green City Rooted in the Land," *Charlotte Sun* (FL), 19 December 2007.

13. "First Babcock Ranch Homes to Be Finished Next Month," *Naples Daily News*, 15 January 2017.

14. "Claiming Murdock Village Is Blighted, Charlotte County Plans to Purchase 3,000 Largely Undeveloped Lots through Condemnation. Is It Constitutional?" *Business Observer* (FL), 10 October 2003.

15. Neil Hughes, "Murdock Village Financing Tops $100 Million, Borrows Tax Dollars," *Charlotte Sun* (FL), 26 May 2007.

# *Index*

Page numbers in *italic* refer to photographs and maps.

MIX
Paper from
responsible sources
FSC® C008955
FSC
www.fsc.org